D0097746

Garden Ponds

Garden Ponds

Basic Pond Setup and Maintenance

Dennis Kelsey-Wood and Tom Barthel

Irvine, California

Karla Austin, *Business Operations Manager*
Nick Clemente, *Special Consultant*
Jarelle S. Stein, *Editor*
Kendra Strey, *Assistant Editor*
Jill Dupont, *Production*
Honey Winters, *Design*
Indexed by Melody Englund

Cover photograph by Adrian Binns, logo photograph (2005©) JupiterImages and its Licensors.
Illustrations are by: **37–39, 63–65, 67–70**: Laurie O'Keefe; **23**: Heather Powers.
The additional photographs in this book are by: **2, 84 top**: Derk R. Kuyper; **3, 7, 10, 87, 92, 98, 104 bottom**: (2005©) JupiterImages and its Licensors; **9, 11–13, 15, 16, 19, 22, 25, 41, 53, 54, 72, 73, 77, 80, 82, 84 bottom, 85, 93, 96, 99 top, 104 top, 105 bottom, 113, 114**: Adrian Binns; **24, 31, 42, 50, 51, 89, 101–103, 105 top:** Sally McCrae Kuyper; **44, 59, 74, 76, 79, 83, 90, 97, 99 bottom, 109–112**: Debbie Dineen/Longshots; **57**: Larry Maupin.

Library of Congress Cataloging-in-Publication Data
Kelsey-Wood, Dennis.
Garden ponds : basic pond setup and maintenance / by Dennis Kelsey-Wood and Tom Barthel.
p. cm.
ISBN 1-931993-69-6
1. Water gardens. I. Barthel, Tom. II. Title.

SB423.K45 2006
635.9'674—dc22
2005033363

BowTie Press®
A Division of BowTie, Inc.
3 Burroughs
Irvine, California 92618p

Printed and bound in Singapore
10 9 8 7 6 5 4 3 2 1

Dedicated to Dennis Kelsey-Wood, whose love of nature and animals continues to inspire readers.

Contents

Introduction: Beauty and Tranquility

The sight and sound of water gently flowing in a pond or watercourse creates a scene of beauty and tranquility, one that can wash away the stresses of daily life. Hospitals and doctors' surgeries feature aquariums to instill a sense of well-being in their patients. Restaurants, hotel lobbies, and many public buildings also use water features to enhance a setting and to create a feeling of peace.

Today, an entire industry has developed to meet the needs of those wishing to feature a garden pond in their immediate surroundings. The types, shapes, sizes, materials, and locations for a pond are many. It is quite possible, and becoming more popular, to own a pond that is actually indoors or located under a covered patio. In Japan and other Asian countries, outdoor ponds may extend into the home, so the owners can enjoy them regardless of the weather. Even if you have only a very small yard, you can plan and develop a magnificent pond to fit it, one to be long enjoyed by family and friends—and you.

In some basic ways, the efficient functioning of a garden pond is similar to that of an indoor aquarium. It differs most in the way the water is contained. With the many choices of materials and equipment now at your disposal, there are endless possibilities. A pond, with its special qualities of beauty and tranquility, is certainly worth the effort required in planning and construction.

Always remember that in the creation of a garden pond, especially one that will support plants and fish, you are creating a self-contained ecosystem. In such a system, everything is dependent on a successful biological balance. Careful planning is essential, and that planning is best achieved through understanding what is required and paying attention to detail.

Garden Ponds is organized to help you understand basic concepts that are the foundation of a successful pond. These concepts are presented first, in this overview, to give you an understanding of just what a garden pond is. Chapters 1–10 provide in-depth coverage of each of the critical issues, as well as technical information that will ensure your success in the planning, construction, implementation, and maintenance phases of your pond.

Basic concepts of the garden pond deal with creating a small, but healthy ecosystem outdoors in water: Just as gardens and aquariums need human intervention to be well established and to

Few things in life can match the beauty of a well-constructed pond that includes hardscapes, colorful landscape plants, and lively pond fish.

remain healthy, so, too, the garden pond requires your knowledge, care, and devotion.

First, you need to understand some key concepts related to a garden pond environment:

- *Water balance.* A healthy pond has a properly balanced chemistry for the fish and plants you want to live there.
    - Acidity and alkalinity are measured as pH values. The proper pH is essential for fish and plant health.
    - Controlling nitrogenous compounds is important to ensure good water quality.
    - Proper concentrations of calcium and magnesium salts provide appropriate water hardness.
    - Water conditioners can remove disinfectants such as chloramine and chlorine that have been added to the water by your water supplier.
- *Oxygen.* Freshwater fish, plants, and helpful microorganisms need plenty of dissolved oxygen.
    - Factors affecting oxygen levels include temperature, daylight and night, and debris in and on the water.
    - Oxygen dissolves in water at the surface, so it is the surface area that determines the size of fish and plant populations.
    - The depth of a pond affects its temperature.
- *Mechanical Aids.* Drainage, aeration, filtration, and pumps are valuable devices that can help you maintain a healthy pond.

Issues related to water quality and maintenance will influence site selection, planning, construction, stocking, and maintenance.

CHAPTER 1

Determining Site, Style, and Size

Having made the decision to feature a pond in your garden or patio, the next thing to do is to consider placement, style, and size of the pond. Many factors determine these, some of which are of a negative type easy to overlook or underrate. Such negatives can turn the project into a disaster or, at the very least, cause the pond to be less efficient and thus create more maintenance work for you. Only when your options are established can you proceed with worthwhile planning.

SITE CONSIDERATIONS

You may already have an idea about where you'd like to establish your garden pond. But does that location meet all the necessary criteria? Here are some major factors to consider.

Legal Aspects

Check local zoning laws governing all aspects of garden ponds. There may be specifications relating to:

- Distance from underground piping (electric, sewage, and water) and surface structures
- The site of an aboveground pond
- Maximum water depth
- Child safety fencing
- Electrical requirements
- Connections to sewer lines

Permits and official inspections may also be needed. It is prudent to check into local codes personally or consult a professional in the area.

Climate

The climate of your region must be considered when planning a pond. Extremes in temperatures and winds will affect your pond. For instance, if the winter temperature dips below freezing for lengthy periods, and if you plan to keep fish and certain plants in the pond during this period, it must be deep enough to ensure the water doesn't freeze to the bottom. To avoid this, a depth of at least 30 inches (93cm) is recommended (preferably more). If you plan only a small and shallow pond, you may need to house your fish indoors during the winter period. Although you can install a pond heater to prevent freezing, this can cause health problems for the fish. (See chapter 10 for more information.)

In addition, cold winds will lower the surface water temperature of small shallow ponds, so you'll need some form of windbreak. It should not be solid; this merely creates turbulence. Far better choices are a trellis-type fence, bushes, small conifer trees (in suitable containers), or a wall of ornamental building blocks. These windbreaks reduce the strength of the wind but do not stop it.

If you live in a region that has hot summers (temperatures exceeding 90 degrees Fahrenheit/32 degrees Celsius), the pond must have some shade during the hottest part of the day. Plants and fish require a minimum of six hours of sunlight a day to thrive, but too much sunlight can raise the water temperature and adversely affect its residents. This will happen if there is no shading and if the pond is too shallow. Fish can be stressed and suffer from sunburn, and their colors can fade.

In certain regions, high winds during the summer can form small whirlwinds that dump sandlike soil into the pond. If this is the case in your area, again, some form of windbreak will be needed for protection.

↑ If your backyard will be shrouded in ice through the winter, your garden pond will need to be deep enough to provide refuge for its fish. When designing your pond, take into account annual temperature extremes.

← For tropical or desert climates, be sure to choose a location with partial shade.

Trees

A beautiful garden featuring a number of mature trees might sound like an idyllic setting for a garden pond. The shade of a tree on hot summer days, as well as protection from inclement weather, make trees and ponds seem a marriage made in heaven. However, you would do well to consider carefully the kinds of trees and their locations.

↑ Trees add dimension to a garden pond but can create nightmare chores when they drop their leaves. If you plan accordingly, even the disaster shown here can be managed with the proper equipment.

For example, if you locate your pond too close to trees, that will increase the amount of maintenance needed, such as removing any leaves that fall into the water before they can decay and cause water-quality problems. This must be done daily, either manually or through use of a skimmer installed in the pond. Small branches that have fallen into the pond can create unwanted vegetation there. Birds perching on branches over the pond will foul the water as will any resins extruded from conifers.

The shade from a mature and spreading tree may reduce the amount of sunlight that reaches the pond. It is also important to learn what type of root system a given tree species has. Some trees have more extensive roots than do others.

The general rule is that a pond site is best no closer to a tree than a distance equal to the height of the tree, or about 50 feet (15m) away, whichever is less.

Utility Services

It's also important to find out where utility lines are located on your property so you do not place your pond over or too close to these. Water pipes, sewer pipes, inspection covers, electricity lines, and any other pipes or cables that may cross the area where you hope to site your pond will usually appear on your property ground plan. Don't hesitate to contact the appropriate utility service authority and have someone come out to mark your property so you'll know what areas to avoid. It's also possible that a previous owner of the property could have installed piping or cabling for which he or she did not obtain appropriate permissions; it's a good idea to have a backup plan in case you have to alter the design of your pond or possibly relocate it.

Be cognizant as well of overhead cables or tall poles that may be located above or near your favored site. When the pool is full of water, you may see their reflections. If you think that will bother you, seek a site where you will see only blue skies reflected.

The final point to consider about utilities is proximity. The farther away your pond is from utilities, the more costly it will be to connect them. When the distance is more than 100 feet (30m), there is a loss in electrical power and water pressure, a loss that increases as the distance increases.

Ground Suitability

You must view the ground of a potential pond site from two perspectives. One is its horizontal suitability: Is it level or sloping? The second is in its vertical suitability: What lies under the surface?

Horizontal. A sloping site has advantages and disadvantages depending on your design. On the beneficial side, you can use the slope to create a waterfall or a two-tier pond arrangement or to create a rock garden or other backdrop to the pond. The major disadvantage is that the slope can flood at its lowest point during heavy rains. You can overcome such a situation by building a partially raised pond. Additionally, if the soil is soft soil, the pond's walls should be reinforced to protect against collapse due to flooding.

Vertical. Because this is the only aspect you cannot actually see, you must undertake a certain amount of work to establish your chosen site's viability. If you're fortunate, the land will be really nice soil, easy to excavate. If the soil is hard, rocky, or sandy, you may have to reconsider the choice of a site. A simple way to learn what the soil conditions are is to dig test holes in various locations. You can do this with a spade, or if you prefer, you can

Your pond site may require extensive excavation. Be sure you are up for the challenge and the ground you have chosen is suitable for digging.

rent a power auger, which will make the process go faster and will be much easier on your back!

POND STYLES

Every pond landscape is a unique expression of the owner's sense of what constitutes beauty, and the potential to be creative is limitless. The aesthetic element of pond design begins with style.

As mentioned earlier, a pond can be either formal or informal, below ground or above. A formal pond has a geometric shape. It forms a rectangle, square, circle, or other shape in which the outline is composed of straight or curved lines, or a combination, resulting in a clearly symmetrical (thus unnatural) shape. In contrast, an informal shape has a very natural look, with little suggestion of symmetry.

Generally, an informal pond is below ground level because this is where you would find a natural pond. It is more complicated, thus more costly, to build an informal pond above the ground and still preserve its natural look. In certain instances, however, landscapers have achieved success with raised informal ponds by placing them at a low point in a raised landscape.

With a formal pond, you have an unnatural shape, which means it will be suitable at or below ground level or partially or totally above. Most ponds seen in the gardens and palaces of the Egyptian, Greek, Roman, and Chinese civilizations were of the formal type because they blended well with the architectural styles of those times. Informal ponds became popular in Japan early on but didn't gain wider popularity until the late eighteenth and early nineteenth centuries, with the development of the classic English Romantic-style gardens. Today, many pond owners utilize Japanese landscaping features.

People who plan to have a pond in a patio or very close to the house often favor the formal pond. If you have only a small yard, a partly or totally raised formal pond can be the best choice. It is easier to build and to maintain. An informal pond begs to have a blending landscape around it, and space may not allow for this unless you're willing to settle for a smaller pond than you may have wanted.

A naturalistic pond such as this one looks as though it has always been a part of the landscape. To achieve this effect, choose your site wisely and take your design cues from Mother Nature.

The strong unnatural lines found in constructions such as this formal pond add a sense of architectural detail to the garden setting.

A fully aboveground pond, however, is problematic in regions prone to subzero temperatures. They will lose heat more rapidly and thus may freeze to the pond floor, unless sufficiently insulated or heated or both.

Although we live in an age in which trends and fashionable thinking change quickly, this is not the case with garden ponds. A traditional, formal style born from centuries of knowledge still holds sway and can prove the safest path to follow when choosing a pond and deciding the most attractive way to landscape it.

However, the present landscape of your garden and the style of your home may call for something less traditional, something informal.

Formal ponds are best suited for the following landscapes and houses:

- The garden design is of the grid style; the lawn has a distinct geometric shape.
- The garden has the look of manicured neatness. Flower beds are organized, with no suggestion that Mother Nature was on their design team.
- The garden features statues, furniture, and other adornments based on classic (such as Greek or Roman) styles and materials. Pathways do not meander.
- The house is of a modern design and features bold geometric shapes.

Informal ponds are best suited for the following landscapes and houses:

- The garden rambles, with a pathway that meanders around a lawn of no particular shape, and bushes, shrubs, and rockeries are placed in a seemingly haphazard manner.
- The garden has a natural country look to it, featuring some wild plant areas, an herb garden, rustic outbuildings, and styled wooden decorations.
- The garden features rocks, country-style furniture of natural wood, and pathways utilizing flagstone, red brick, or similar natural stonework.
- The house has an older look and may feature exposed beams in the Tudor style or has ivy, clematis, or other climbing plants adorning its brickwork or on nearby walls.

A garden landscape may have elements that favor both formal and informal pond styles. In this case, you need to consider your proposed site, its landscape elements, and which of these you could easily change to better suit the pond style you prefer. A garden pond project may be short term in its establishment, but the pond may become a long-term project when considered as part of the total landscape. It is best to consider the whole and think long term as you may want to modify the entire scenic view to better suit your preference for a given pond style.

POND SIZE

In determining pond size, the rule of thumb is that the pond should be neither too large nor too small for the available space. For instance, while you can use an internal or under-gravel filter in a small pond, a large pond will require space for an external filter system. If the site for the pond is small, however, you must allow for room to move around the pond to attend to its maintenance. If the pond size you would like would prevent or hinder this, it is wise to reduce it.

Size also depends on whether you wish the pond to be the main focal point of your garden or merely one of many features in a landscape containing numerous attractions. In the latter case, you will want not a grand pond but rather a small one that provides an interesting spot where you and your friends can stop awhile to chat and see the fish, wildlife, and plants on view in your watery oasis.

Bear in mind as well that the size of the pond will determine the population of fish. This is because the surface area, not its depth, controls the amount of oxygen the water can absorb. This in turn controls the number or length of the fish that can survive in the pond without the need to supply them with additional oxygen by mechanical means. We will discuss this aspect in greater detail in chapter 4.

Finally, the amount of cash available for all pond and landscaping costs will determine just how ambitious you can be. It is important not to compromise quality in your pond decision making. It is far better to scale back on size than to invest in inferior materials or operating systems (pumps, filters, drains). Make sure the entire pond project is made of high-quality material, thus ensuring longevity of wear. This means you may have to complete other aspects of scenic landscaping in stages when funds are available. Nothing is more annoying than encountering problems, such as pond leaks or inefficient filters, just when you thought you had completed the project!

➊ A long, narrow pond is suitable for this small yard.

CHAPTER 2

Pond Materials

Installing a garden pond has never been easier, thanks to the many options for materials, equipment, and accessories now available. The range is wide enough to satisfy all tastes and budgets. Before deciding on a direction, shop around to see how extensive the garden pond range of products is. Collect catalogs and give yourself plenty of time before settling on which materials, equipment, and accessories will be best for you.

PREFORMED PONDS

The preformed or molded pond, also known as a rigid liner, is the most popular for smaller ponds. It is the easiest (thus quickest) to install, has a long service life (twenty to fifty-plus years depending on material), and is puncture proof. It is also available in a range of styles and sizes, a number of colors, and two texture finishes—smooth or uneven. The material used in its manufacture is either rigid plastic (high-density polyethylene at the top end of the range, with a twenty-plus-year life) or fiberglass (a minimum of 0.25in [6.3mm] thick for the best quality and with a fifty-plus-year life).

Another advantage of quality preformed ponds is that they are more resistant to high and low temperatures as well as ultraviolet rays, which will cause lesser quality materials to degenerate. Their sloping sides further help them to resist freezing to the bottom of

➲ Using a preformed plastic tub, such as the one shown here beside a flexible liner, is the quickest and easiest way to install a small garden pond.

the pond. Finally, it is easy to move a molded pond from one potential site to another, allowing you to get an idea of how it will look in various locations.

One disadvantage is that preformed ponds are rather limited in their upper sizes in respect to both surface area and depth. However, there are companies that specialize in custom-made moldings (though these tend to be expensive). There are companies as well that produce these ponds in sections you can bolt together and seal with marine silicon so they are fully waterproof. You can also select from a variety of waterfalls (made of the same materials), which you can add to create a very natural-looking combination.

A preformed pond can be either formal or informal in style, and most preformed ponds have one or more shelves on which you can set aquatic plants in pots. If you wish to lure amphibians to your pond, as well as to feature marginal plants (which grow well in partially submerged conditions), there are preformed ponds with shallow wide ledges for this purpose. You need not worry about calculating the volume of water needed to fill a molded pond because the manufacturers provide details about this in their specification labels or leaflets. Table 1 provides you with a sampling of popular sizes of ponds in the range of 72–1,200 gallons. All include shelves for plants.

TABLE 1 SAMPLES OF PREFERRED POND SIZES AND VOLUMES

Length Max.	Width Max.	Depth	Volume (US)	Style
5ft (1.5m)	4ft 9in (1.4m)	20in (51cm)	72 gal (273 ltr)	Formal
6ft 9in (2.0m)	5ft 4in (1.6m)	24in (61cm)	168 gal (637 ltr)	Informal
8ft (2.4m)	6ft 9in (2.0m)	24in (61cm)	252 gal (955 ltr)	Formal
9ft 3in (2.8m)	7ft 4in (2.2m)	24in (61cm)	372 gal (1,410 ltr)	Informal
10ft (3.0m)	6ft 9in (2.1m)	26in (66cm)	504 gal (1,915 ltr)	Informal
11ft 9in (3.6m)	8ft 9in (2.6m)	30in (77cm)	780 gal (2,957 ltr)	Informal
16ft (4.9m)	7ft 3in (2.2m)	30in (77cm)	1,200 gal (4,550 ltr)	Formal

Always remember when viewing preformed ponds that the larger ones look enormous when freestanding but appear to shrink dramatically once installed in your garden. To get a real idea of size, try to visit a garden center that has examples in the ground.

A final point to consider is that some preformed ponds have a convex edging while others are flat or rippled. This aspect has importance when you are choosing how to conceal the pond edging. As this type of edging creates an uneven surface, you may need to hide the edge with creeping marginal plantings or by building up level surfaces behind the edge, on top of which can be placed slabs or patio bricks. If you are using slab or brick, be sure to protect your pond material edging so it will withstand the weight.

With respect to color, black and dark green are usually the best. They give the pond the appearance of being deeper than it actually is. They also create a good background to show off the colors of koi and other pond fish. Avoid blue and aquamarine as they look very artificial.

FLEXIBLE POND LINERS

Although more work intensive to install, a flexible liner has many advantages, the most important being that it is not size limited. It is the choice for the more serious enthusiast who requires a large pond, or who needs one with good depth to house a koi collection, or both. The flexible liner enables you to design your own pond shape and its depths. If you want a very large pond, you can bond or seam various liner lengths together. Seaming kits are commercially available and include waterproof epoxy adhesives that are used to permanently seal two lengths of flexible liner. These liners are suitable for any pond style. If you desire a formal pond of a simple shape, the liner manufacturer will box-weld this for you so it is easier to install.

An obvious danger to flexible liners, even quality thick ones, are sharp objects in the ground, which can puncture a liner once the massive weight of water fills the pond. To avoid this danger, you must make sure that the pond site is smooth and free of anything that might cause damage. In addition, it is prudent to have an underliner (underlayment). This is available in various thicknesses. In some instances, liner producers fit their liners with a protective material. If a liner doesn't have one, sand is fine for the pond base, and old carpeting or similar material or cardboard will work for the sides.

Ultraviolet (UV) light can be another problem; it can degrade low-cost materials, making them brittle and easy to tear. Fortunately, many liners today, especially the more expensive ones, are resistant to UV light. The latest liner styles come com-

➔ Flexible liners offer maximum versatility and can be cut to fit virtually any shape or size garden pond.

plete with either gravel or small pebbles embedded in them, giving the pond extra protection from UV light as well as a natural look. If you're unsure whether your liner is resistant or you just want to make sure no damage occurs, you should protect any exposed liner edge and sidewalls with slabs or other materials that overhang the water. Bear in mind that evaporation of the water lowers its level, so the overhang should allow for this.

Liners come in various materials, each offering advantages (service life, cost, strength) over its alternatives. The following is a summary of liner materials. *Note*: Whichever you choose, check that the material is safe for both fish and plants. For instance, PVC produced for swimming pools, roofs, or other construction trades may be toxic to fish, as it contains harsh industrial fungicides and UV light protectors that can leach into pond water. Be sure the materials you choose are free from these additives.

Polyvinyl Chloride (PVC)

With a life of at least ten years or more (depending on thickness), PVC is the least costly of the pond liners. It is lightweight and available in various thicknesses—0.02in (0.5mm) to 0.05in (1.2mm) being the popular range. You can also buy PVC-E (meaning enhanced), in which the basic material is strengthened by woven strands of nylon. It has good UV light resistance and retains flexibility during cold weather.

Ethylene-Propylene-Diene-Monomer (EPDM) and Butyl

Both of these materials are rubber based and available in a variety of thicknesses beginning at 0.02in (0.5mm); the high-quality end of the range is about 0.04in (1mm). A number of these liners carry a lifetime warranty, others twenty-five years. Their only real disadvantage is their cost—a minimum of 10 percent more than PVC per square foot.

Xavan

This patented product is made of polypropylene and is presently the most expensive of the liner options. It has very high resistance to UV light and is about one-third the weight of EPDM or butyl and is three times more tear resistant than either of them. It is flexible to -70°F (-21°C), and its light weight makes it easy to work with.

LINER SIZE CALCULATIONS

For a simple formal shape the calculation is:
Length + twice depth + twice edging
Width + twice depth + twice edging

Example
A 20 x 10ft pond with depth of 4ft and 1ft edging
Liner length needed 20 + 8 + 2ft = 30ft
x = 600 sq.ft (55.7 sq.m)
Liner width needed 10 + 8 + 2ft = 20ft

If an island is to be part of the pond, add extra liner to allow for this. For informal and circular ponds the liner size will be that rectangle or square into which the pond will fit. This will result in some wastage, but it may be possible to use this for other purposes.

CONCRETE PONDS

Prior to the 1950s, which is when pond liners and preformed ponds came on the market, concrete or tiled brick was the natural choice for a pond builder. The major problem with concrete is that it requires a degree of skill to work with and is prone to cracking, especially in regions in which subzero winter temperatures are normal. A further problem for the average pond owner is that the concrete pond requires the building of timber forms to hold the concrete while it hardens. For these reasons, it lost popularity as technological progress was made in the various forms of synthetic compound liners and preformed fiberglass ponds. However, developments in concrete still make this material an option either alone or in combination with a liner or fiberglass.

The addition of plastic fiber to the concrete mix results in a very strong material, one resistant to water and cracking. Its wear life is excellent and greater than that of its rivals. You can overcome the need for wooden forms by having a contractor spray Gunite (shotcrete) concrete onto a wire mesh–lined excavation. In the case of the formal pond, concrete blocks strengthened with rebar and placed on a solid concrete base will provide the framework that avoids wooden forms. A cement/fiber mix then seals the concrete blocks. A liner will achieve the same objective, as will a coating of fiberglass. However, this all adds up to more work and more cost. For anyone not having excellent do-it-yourself capabilities, the preformed pond or the pond liner probably is the better option.

FIBERGLASS PONDS

When it comes to an ideal material for pond making, there is little doubt that fiberglass is the answer. It is immensely strong, lasts a long while, repairs easily, and can be configured readily into any shape without the unsightly folds, and their associated problems, that can occur with liners. Its major drawback is that it is extremely messy to apply and very costly in comparison with the other materials discussed. Definitely not a material the unskilled should play with! It is possible to spray fiberglass directly onto a suitable excavation, and a growing number of companies specialize in this method. But again, this is costly. Fiberglass is therefore best in the form of the preformed pond for the average novice pond builder.

CHAPTER 3

The World of Water

Water is the elixir of life because without it neither plant nor animal can survive. When you look at a beautifully landscaped garden filled with plants and delightfully colored fish, it is easy to overlook the fact that the entire vista is artificial. What you are seeing is a combination of considerable artistic flair and a world of water in balance with nature. Chances are that the pond receives a high degree of mechanical help to ensure that the water remains clean and healthy.

Left to its own devices, a pond of static water soon becomes stagnant. It is then a body of foul-smelling green liquid, an algal paradise devoid of fish (they all expired), supporting plants on the basis of survival of the fittest, which usually means the weeds win out. Not uncommonly, such ponds indicate that the owner had the desire and willingness to create a watery world but failed to understand the principles on which its success is so dependent.

It is important to remember that when things go wrong in a pond it is usually due to a number of factors, rather than only one. Bear in mind, too, that "quick fix" treatments to cure a problem will, at best, do so only temporarily. If the underlying cause of the problem remains, then the problem will reoccur.

➔ Maintaining a clear, healthy pond requires a basic knowledge of chemistry and constant help from mechanical components.

Let's briefly discuss various aspects of a pond so you can avoid some problems or at the least be aware of how to overcome them if they do manifest themselves. This knowledge will enable you to obtain the right equipment (such as filters and pump) and a sensible balance of fish and plants to ensure your pond and its inhabitants live in harmony. We'll discuss certain topics at greater length in the chapters that follow.

WATER CHEMISTRY FACTORS

A pond is a biological system that is never static but forever changing because of the countless biological and chemical interactions continuously taking place. Therefore, imbalances can arise, and if not attended to, minor problems may rapidly escalate, negatively affecting the desired efficiency of your mini-ecosystem. It is important to understand the basic factors of water chemistry.

Acidity or Alkalinity (pH Values)

All water organisms, including plants, have evolved to live in a preferred range of acidity or alkalinity of water, the pH range. Some are able to tolerate a wider range than others so it is important to know the range for any fish, invertebrates, or plants that are to be a feature of the pond.

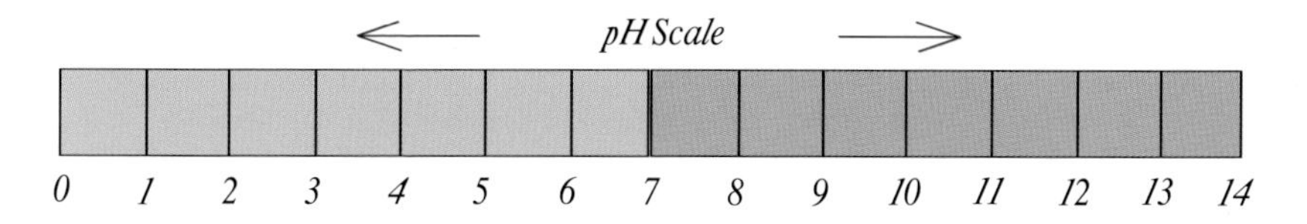

The pH value of water or other aqueous solution is a logarithmic index of the degree of acidity or alkalinity of the solution. The scale indicates the concentration of hydrogen ions in relation to hydroxyl ions. If these two differing ions are equal, the water is neutral. If there are more hydrogen ions, the water is acidic; if there are more hydroxyl ions, it is alkaline. The index operates from 0–14. At 7 the water is neutral. Below 7 it is acidic, and above 7 it is alkaline. Each unit of change represents a tenfold movement, which means a reading of pH6 is ten times more acidic than one of pH7. If the water is at pH5, this is one hundred times more acidic than at pH7. The same is true of movements above pH7.

➊ A pH of 7 indicates a neutral environment; below 7 indicates acidic water while above 7 indicates alkaline. Each unit of change represents a tenfold increase or decrease.

Most pond fish and plants are happy with a pH of 7–7.5 but can tolerate short periods when it falls to 6.5 or rises to 8. Indeed, during any twenty-four-hour period, the pH will fluctuate and can even reach pH11 when there is strong algal bloom.

➔ Commercially available products such as these can be found in most pond supply stores and help balance water chemistry when needed.

But this should last only a short period before falling back to the desired reading.

The most commonly kept pond fish include koi and a wide range of goldfish (such as comet, ryukin, and shubunkin), which generally thrive in a pH of around 7.0 to 8.5. Other fish, more tolerant of swings in water quality and pH, might include native game fish (such as black crappie, sunfish, bream, catfish, and tilapia), which are sometimes placed in garden ponds. Other hearty fish that consume mosquito larvae include killifish and gambusia—sometimes available through your local vector control department, free of charge. In warmer, subtropical to tropical climates, hobbyists have experimented with keeping aquarium species, such as plecostamus, mollies, and oscars, which must be brought indoors when weather conditions turn unfavorable.

You can measure the pH by adding a reagent from a test kit to a sample of the water. Once shaken, the solution will change color and can be compared with a color chart supplied with the kit. This will indicate the pH value. You can also purchase an electronic pH meter. You can also purchase tests strips made for ponds or pools. It is prudent to check the pH on a regular basis—the more so if the pond is small.

If pH testing shows an imbalance in the water chemistry of your garden pond, there are several steps you can take to correct it and bring the levels back to acceptable levels. Start by increasing aeration and performing daily water change outs. Begin by removing 10 to 25 percent of the water and replenishing it with dechlorinated, pH neutral water. Test the pH after each change out to monitor any improvements. Measure the pH again after each twenty-four-hour period. If greater swings (two degrees or more) in pH are detected, start with a change out of 25 to 50 percent of the water. If swings of 3 or more degrees in pH are detected, remove the remaining pond fish immediately. Specially formulated acid buffer and pH increaser products can also be added and are designed to adjust mild swings in pH. They should only be added if water changes fail to correct the problem.

Other, harsher chemicals can be damaging to you, your pond fish, and the beneficial bacteria that help keep the garden pond in balance. They should only be used under emergency circumstances and with caution in small, dilute amounts. For ponds with high alkalinity, remove the fish to a balanced environment and add small, dilute quantities of muriatic acid (formulated for use in swimming pools). Allow the water to circulate fully before testing for correction; add as needed to bring the water back into balance. For high-acid water, add calcium carbonate or materials containing the same chemical formulation (concrete blocks [not cinder], pulverized oyster shell or limestone gravel). Do not use common, slaked lime marketed for use on lawns, as it can cause toxic reactions in pond fish.

Nitrogenous Compounds

There are three nitrogenous compounds of importance to the pond owner: ammonia (NH_3), nitrite (NO_2), and nitrate (NO_3). Their levels of concentration are indicators of the water quality. For ammonia, as a general guideline, at a water temperature of 70 degrees and a pH of 7.0, acceptable levels would be 1 part per million (ppm). At a higher pH of 8.0, as little as 0.1ppm could prove toxic. Nitrite is dangerous at concentrations as low as 0.25. At even lower concentrations, over extended periods of time, it can cause adverse health conditions in pond fish. For nitrate levels, readings from zero to 200ppm are considered acceptable.

Ammonia is a highly toxic gas created as a by-product during the decomposition of organic waste, such as dead leaves, decaying

← Algae blooms such as this one that cover the entire surface of the pond are notoriously difficult to control and occur when nitrate levels are elevated and direct sunlight is present.

organisms, uneaten food as well as fecal matter resulting from fish's and other animals' metabolism. Even at low levels, ammonia will adversely affect the health of fish—the smaller ones may die. As the concentration increases, the larger fish will suffer and may die.

Ammonia is oxidized by *Nitrosomonas* bacteria. Nitrites are almost as dangerous as ammonia. Nitrite is sometimes referred to as the silent killer, because it can cause organ malfunction and death. Another bacterium, *Nitrobacter*, in turn converts nitrites to nitrates, which are far less dangerous to the fish. Plants absorb nitrates as part of their nutritional needs, thus completing what we call the nitrogen cycle.

However, if the nitrate concentration exceeds 200ppm, plants and algae begin to grow rapidly. As the pond fish consume this plant material, they produce excess excrement, and the whole system is thrown off balance by higher levels of ammonia and nitrite. This ever-changing nitrogen cycle must be monitored and managed.

To overcome the problem of high ammonia content, the immediate remedy is to replace about 20 percent of the pond water. You may need to repeat this for two or more days until you obtain the required reading of zero or only a minute trace (under 0.1mg/liter) of ammonia. It is best to draw the water from the lowest point in the pond because this has the least oxygen and therefore the greatest concentration of decaying matter, thus the most ammonia and nitrites.

If the pH reading is neutral or acidic, this lowers the ammonia present (good news) but increases the nitrite content (bad news). The bottom line: you need to make regular checks for these compounds with easy-to-use test kits. For the desired bacteria to proliferate, they require well-oxygenated water.

Once you achieve normality, it is essential to review all aspects of husbandry to overcome the causal problem(s). With a small pond, say under 500 gallons, it is wise to replace about 10 percent of the water every week during the warmer months as this will help to prevent a steady buildup of toxins and should help keep oxygen levels healthy.

Calcium and Magnesium Salts

During their passage over various rocks and soil types, natural waters absorb many mineral elements and compounds along with organic substances. Their collective effect is to make the water either soft or hard, which refers primarily to the amount of calcium and magnesium salts in the water. The higher the levels of both salts, the harder the water. This is expressed in either German degrees (odH) or parts per million (ppm) of calcium carbonate.

Hard water is beneficial to freshwater fish, such as goldfish and koi, because it reduces the workload of the osmoregulation system that operates to maintain cellular balance between salt and water.

For instance, if the water is soft and contains low levels of dissolved salts, it is out of balance with the fish's salty body chemistry. The fish will be forced to take in excess water through its skin by osmosis to dilute its own salts and achieve equilibrium with the surrounding water. As the fish takes in this excess water, its body swells. In order to expel the excess water, the fish produces large amounts of urine.

Most pond fish prefer slightly alkaline waters, which fortunately are normally moderately hard with a odH of 6–12. This equates to 107–215 ppm. (*Note*: To convert odH to ppm, multiply by 17.9. To convert ppm to odH, divide by 17.9.)

You can purchase test kits to measure water hardness. If testing results indicate that the pond water is below the desired hardness, you can add a calcium-based material, such as crushed oyster shell, to one of the filter chambers. If your water is too hard, you can add a calcium or magnesium buffer product, commonly found at pond supply stores.

Chlorine and Chloramine

To make our domestic water system safe from unwanted pathogens (disease-causing organisms), water authorities add disinfectants to it. Chlorine, a toxic gas, is often the favored choice. The amount added (0.1–0.2mg/liter) does not harm us, but it can kill pond fish. Fortunately, chlorine is not very stable in water and rapidly dissipates at the surface. The same is not true of chloramines that water suppliers also use, which take much longer to remove without chemical additives.

When initially setting up your pond, pass the water through the filters for a few days (and before you introduce any fish) as this will remove the chlorine. You can obtain conditioners for the water that will remove chloramines (as well as chlorine) in a relatively short period.

During the summer, water authorities may increase the amount of disinfectants used, as well as flush their systems with other chemicals to remove parasites. During this period, monitor your water quality carefully. You may also wish to ask your local water board what disinfectants they add.

Oxygen (O_2)

If there is one aspect of maintaining a garden pond that every owner must appreciate, it is the importance of having adequate quantities of dissolved oxygen in the water. It is also, without a doubt, the most abused aspect of aquatic management and probably accounts for the death, directly or indirectly, of millions of pond and aquarium fish every year. When the concentration of oxygen falls below certain levels, there is a rapid deterioration of the water quality and a negative effect on the fish as well as on many other elements. Fish are not the only consumers of oxygen. Plants need it when they respire, as do thousands of small and microscopic life forms and bacteria that also dwell in the water and form an integral part of its quality.

There is a correlation between water temperature and oxygen. As the temperature rises, the oxygen content falls. As water temperature increases, so does the physical activity of the fish and other organisms, thus creating an increase in their oxygen consumption. A low oxygen content also increases the potential for normally nonpoisonous compounds to become toxic.

Pond fish require water with a minimum of 6mg/liter of oxygen, and stocking rates (see chapter 9) are generally calculated on the basis of a 68°F (20°C) temperature. Table 2 indicates the effect of temperature on oxygen content.

TABLE 2 EFFECT OF TEMPERATURE ON OXYGEN CONTENT

Temperature	Mg/Liter	Percent of Oxygen Reduction
41°F (5°C)	12.4	
59°F (15°C)	9.8	21
77°F (25°C)	8.1	35

The term *balance* means exactly that. On one side of the scale you have the total oxygen content of the pond water and on the other the total consumption of that oxygen throughout the day by the fish, plants, and microorganisms living in that water. If consumption is greater than content, there is no balance. If the content is always balanced it is fine to have a surplus of oxygen, as this will allow for plant and fish growth or allow you to increase the stocking level somewhat if you so desire. Clearly, it is prudent to test the water for its oxygen level, which you can do using either a test kit or an electric meter. To establish the range of oxygen content, test the water during the late afternoon on a sunny day and then the following morning around 6 or 7 a.m. This will give you a reading range from a point at which the O_2 content is at its highest to one at its lowest. The overnight drop in oxygen content happens because during the day the plants produce oxygen as a result of photosynthesis. At night oxygen production ceases, and the plants consume oxygen—in large quantities if the pond has many plants.

The low point is the critical one because if it is below the needed level for the fish, they will stress. If it is near zero, they will gasp at the surface for air. The next step is death from suffocation.

Once a problem is manifest, the first corrective step is to increase aeration and filtration rates, then drain some water from the low point of the pond. This will increase the water circulation rate and draw oxygen to the lower levels of the pond. Next, remove all debris from the pond, especially any floating on the surface. Check that the filter or filters are clean and free of clogging.

Consider whether you have too many plants with leaves on the surface (thus reducing the interface at which oxygenation takes place) or too many fish—especially large ones. Be as sure as you can be that you are not overfeeding the fish because uneaten food will decay—and in the process consume oxygen.

Keep in mind that even a well-filtered and ventilated (aerated) pond may suffer from oxygen shortage.

POND SURFACE AREA

Contrary to what many beginners think, the number of fish inches that can survive in a given body of water is determined, not by its volume, but by its surface area. This is because oxygen dissolves in the water at its surface. The larger this area, the greater the number of fish inches the pond can sustain. Aquarists talk of fish inches rather than fish per se in respect to stocking levels because the two are quite different. Fish come in many sizes, and their oxygen needs are in ratio to their size. A big fish consumes more than a small one does.

If a given pond can support 100 fish inches (254 cm), this means a hundred 1-inch (2.54 cm) fish, ten 10-inch (25.4 cm) fish, or any other combination of length and number that adds up to no more than 100 inches. Only the body of the fish counts—you can ignore its tail length. Ambient temperature, number of plants, hours of sunshine, and whether mechanical aids to increase the oxygen are in use will all influence the amount of oxygen content. However, your base guide is the water surface area. To calculate this, multiply the pond length by its width. For example, a 15 x 10ft (4.6 x 3m) pond has a surface area of 150ft^2 (13.8m^2).

If the shape is informal then geometrically divide this into two or more convenient formal shapes, calculate each of these, then add the figures to produce an approximate surface area. If you have a circular pond, multiply the radius by itself, then multiply the result by 3.14. For example, if the pond had a diameter of 20ft (6.1m) then its radius is 10ft (3m), so 10 x10 x 3.14 = 314ft^2 (29m^2).

POND VOLUME

You need to know the volume of a potential pond so you can purchase a pump of appropriate power to move the water through the filter system—assuming you plan to install a filter, which is strongly recommended. Knowing the volume is also necessary if you ever need to add medicaments or other treatments to the pond. To calculate the volume, multiply length by width by depth.

For example, if a pond is 15 x 10 x 4ft (4.6 x 3 x 1.2m), its volume is 6,300ft (16.63m).

A cubic foot holds 7.48 US gallons of water so the sample pond holds 600 x 7.48 = 4,488 gals. This equates to 17,000 liters, or 3,738 UK gals. See the Appendix for some useful conversions so you can calculate any volume in either metric or nonmetric units, whichever you are more comfortable using or need.

The depth of a pond is important in that it affects the water temperature. A deep body of water—more than 4ft (1.2m)—remains stabler than a shallow one. It is also better for the fish psychologically and physically, allowing them room to exercise. During the winter period, a deep pond offers the fish a haven in its lower levels when the higher levels are frozen.

➔ A pond's surface area determines the number of fish inches the pond can accommodate.

CHAPTER 4

Drainage, Aeration, Filtration, and Pumps

The continued success of any garden pond rests on whether the owner ensures from the outset that the water is reasonably clear, free of offensive odors, and capable of sustaining plants, fish, and other organisms in a healthy state. The best way to achieve this is by the use of mechanical aids such as drains, aerators, filtration systems, and pumps.

Should you choose to install one, bottom drains are located at the pond's lowest point. They are used to remove water containing detritus and foul biological sludge. As this sludge and waste collect, they can be periodically flushed and expelled at a distance from the pond, adding nutrients to flower beds or vegetable gardens while keeping harmful chemicals out of the garden pond.

There are numerous ways you can aerate and filter your pond according to its size and your expectations of it, be this a small general pond or a large and magnificent setup to house a collection of multicolored koi. The differences between the extremes of the small and the large are not those of principle but of extent. Here we will discuss the fundamentals of aeration and filtration so that armed with this knowledge you will be able to consider your options and determine what system is best for your particular pond.

Although in real terms the filtration system will normally encompass aeration (oxygenation or ventilation) of a pond as part of its function, we will discuss aeration and filtration individually. This will enable the first-time pond owner to appreciate that aeration and filtration are separate processes that you will normally integrate in order to share the same pipe work. (*Note*: This may not be so in large pond setups featuring waterfalls and fountains.)

But when we're talking about basic, small, beginner garden ponds, filtration and aeration are usually combined. Single units typically perform both functions, as in the case of a filter falls unit. The filter cleans the water and also aerates it by trickling it downward, over rocks, to introduce oxygen to the water. Sometimes, the filter box also uses what's called a "venturi" device, which injects oxygen into the water before it returns to the pond. In larger ponds, however, this sort of approach does not supply the needed oxygen. You must introduce added aeration through the use of a fountain or separate aeration pump.

AERATION

The object of aeration is to take the water from the pond, pass it through a pipe, and return it at the opposite end of the pond in such a way that it will agitate the water. This increases the surface area of the flowing water, exposing it to atmospheric oxygen. This oxygen is then dissolved into the liquid water as it returns to the pond. At the same time the circulation this creates will minimize the potential for temperature stratification and the retention of low oxygen content at the lower levels. The return pipe of the water also can feed a waterfall, a fountain, or a spray pipe (see chapter 5). In each of these accessories the exposure of the water to air further increases oxygenation.

Another means of increasing aeration is to fit a venturi pipe to any point in the pond's circulatory system before the water returns to the pond. To create a venturi pipe, a narrower, restricted pipe is fitted to the existing pipe. This venturi pipe is then fitted with an aperture that you can open or close in order to adjust the amount of air flowing into the venturi. When water enters the restricted return pipe, it accelerates and creates a partial vacuum at the venturi that sucks air down through exit holes along the length of the venturi pipe, thereby introducing increased oxygen back into the pond. Water from the venturi returns to the pond just beneath the water surface, sending up air bubbles that burst at the surface, increasing surface area, agitating the water, and oxygenating it. The venturi then exits into the last piece of outlet piping just beneath the water surface, sending up air bubbles, which burst at the surface, increasing surface area, agitating the water, and oxygenating the water. You can place venturi pipes at any point in the circulatory system, and you may see them at appropriate points of biological filter systems where there is a need for extra oxygen. Venturi devices are often found integrated into filtration systems in a special chamber, either right before or right after the water passes through biological filtration media.

Circulating the water and returning it to the pond oxygenated will not of course remove any of the unwanted gases, nitrites, or detritus that make it unhealthy. For this, you need to insert a filter system into the circulation. You also need a pump to actually move the water around the system, but we will look at this later as your pond size and accessories (such as filter, fountains, waterfalls) will determine the type and size of pump to use.

FILTRATION

The objective of any filtration system is to take in dirty and unhealthy water and convert it to healthy water. If, as a consequence, it also makes the water very clear, that is a bonus, but it is not the main objective of good filtration (as many novice pond owners think). Clear water may contain many toxins in a dissolved state, thus not visible to the eye. It is therefore beneficial to understand how a good filter system works to make the water healthy before all else.

Total filtration systems, such as those used in most backyard garden ponds, function on two (and sometimes three) basic principles—mechanical, biological, and to a lesser degree, chemical. The first component of a total system is mechanical filtration, which removes large solids and waste from the water. The water then passes through a biological process, whereby filtration media, colonized by beneficial bacteria, come into contact with the water. The beneficial bacteria consume the harmful ammonia and nitrite toxins, converting them to nitrates, which benefit garden pond plants. Each of these processes is performed in a single chamber or a series of chambers, which are, in turn, cleaned and maintained by the pond keeper. Older filtration products contain a third medium, used to remove harmful chemicals.

Always remember that you can never have too much filtration, but you can certainly have insufficient filtration to meet the needs of plant, fish, and other life forms in the pond. If you are not building your own filters, it is best to buy well-established brand name systems.

Bear in mind also that, even if you have an excellent filter system, you must not let it overload because of your desire to increase the stocking rate beyond what it was originally or because you did not take into account the ultimate growth size of the fish. This is especially true with respect to koi.

For our purposes we will discuss a filter that comprises a number of chambers and a watercourse. This is what a serious hobbyist will use if he or she has a large pond. However, in a smaller pond you can have only two or three chambers (or even a single one) that contain the three components of filtration.

DRAINAGE

Small ponds, and even larger ones with only a small number of fish, do not necessarily need a bottom drain because these are costly to fit and require more spadework to provide a channel for the piping. The use of pond vacuums will be quite adequate for the average garden pond.

However, in a large pond it is beneficial to have a means of removing the dirtiest water at the bottom of the pond so there is less strain on the filter system. You can achieve this with one or more drains fitted at the lowest point of the pond. It is important to place the drain in a concave at the pond base to draw sludge and dirty water when the drain is open. A pipe feeds the water to the sewer system or to other means of disposing of it. (*Note*: If flushing it into the sewer, ask your water authority if there are any regulations that place limitations on this.)

A second type of drain you may require is one that carries potential overflow water from the pond following heavy rain. This (or these) you can install just above the high water mark of the pond so that any extra water will enter the overflow and channel to the sewer system, or to any other point that is below the level of the overflow inlet pipe. Be sure to protect the overflow inlet with a mesh of a size small enough that it will not suck small fish into it!

Methods of Filtration

In actuality a filter system does more than simple filtration because part of its function is to enable biological conversion of toxic compounds into safe compounds. It will achieve its full function of mechanical filtration and biological filtration. Here we will consider them as individual components, though.

Mechanical filtration encompasses the preliminary removal of heavy detritus by gravity. For large-scale ponds, this is typically accomplished by using one or more settlement tanks, which allow heavy detritus to sink to the bottom and the water to pass along to the next stage of filtration. Water passes from the pond into a settlement tank, where the heaviest (and usually largest) pollutants sink to the bottom. The settlement tank contains a bottom drain that removes heavy sediment once its outlet pipe gate valve is open. The insertion of baffle plates will slow the flow of water and allow more of the smaller detritus to fall to the tank floor, enabling the tank to function at its best. In a multichamber system the water will enter the settlement tank near the bottom and exit at the top.

The mechanical filter appears in the second tank in the form of nylon brushes or other inert material, such as large gravel, hair rollers, mesh, foam material, special glass, or plastic pieces of tubing—even domestic pot scourers. Each of these materials will prevent detritus of varying sizes from onward movement through the system. Ideally, two or more types of filtration materials will be used, the last of which should be foam. This will trap the smallest of particles. Whenever they are obviously full, remove and clean the mechanical filters.

In smaller self-contained systems, water passes through a series of small chambers containing baskets with different sized openings, much like skimmer baskets found in swimming pools. The openings are progressively smaller, removing all but the smallest particles. Even smaller particles are removed with a series of foam pads of varying densities. The water then travels on to the biological stage.

Biological filtration revolves around the nitrogen cycle and the bacteria discussed in the previous chapter. These bacteria are present in the atmosphere and in water. Within an oxygen-rich environment that has suitable surfaces (media) to live on, they will form a thriving colony. You can jump start a biological filter by adding freeze-dried or live cultures from your local garden center. Add these to the biofilter chamber, and they will get the biological process underway.

Any of the materials in use for mechanical filtration will serve as biological filters as well. Those such as nylon brushes and gravel have large surface areas that make them ideal for bacterial colonization. Smooth pebble-type gravel is the least effective because it has a much smaller surface area.

A biofilter will normally be placed after the mechanical filter so that the water is reasonably clean as it passes over the biomaterial surface. Of course, if the settlement tank contains mechanical filters, these also will serve a biological function.

When cleaning the biochamber, clean about one-third at a time. Additionally, use pond water to clean the filter because tap water will contain chlorine and other disinfectants that kill the beneficial bacteria, which would cause the filter to lose its effectiveness and require weeks, or even months, to recolonize. Only clean the filter when debris is clogging it to the degree that the flow rate is impaired.

Once the water leaves the biological filter chamber it will still contain nitrates, and these the pond plants will absorb as nutrients. However, if a pond contains no plants, or only a few, then the addition of either a watercourse or any type of open chamber stocked with a plant such as watercress will remove many of these.

Chemical filtration involves the removal of ammonia and nitrites from the pond water, whereby these compounds form loose bonds with appropriate filtration media. In the past this was often activated charcoal or zeolite. They were a feature of nonbiological filters. However, this form of chemical removal of ammonia and nitrites is no longer necessary given the availability of high-performance biological filter systems that do a better job with fewer associated problems. I mention this form of filtration purely because some owners still use it.

Filtration Systems

Filter systems are available in four forms: gravity-return, pressure filter, gravity-fed multichamber filter, undergravel filter.

A *gravity-return filter* uses an in-pond pump to push the water to the filter chamber situated above the water surface. The water passes through the filter and exits at the pond surface from an outlet pipe near the bottom of the filter.

A *pressure filter*, popular for use in the smaller pond, also uses an in-pond pump. However, unlike the gravity filter, it is a sealed unit within a casing, usually a cylinder shape, which is under pressure. As a consequence the water is still under pressure when it leaves the filter so the unit need not be above the water surface level and therefore is easy to conceal in the ground or behind a bush near the pond.

The *gravity-fed multichamber filter* is popular for use with large ponds as well as with koi devotees. For this type of system to succeed, its water surface must be level with that of the pond water. Water flows by gravity into the bottom of the first chamber and exits into the next chamber at the top over a transfer port. It then flows down into a void area and up through a filter medium supported on a perforated tray to exit over the next transfer port. This repeats until the end of the filter system, which may contain two or more chambers.

In this system the pump pulls rather than pushes the water so it is normal to site this at the end of the system in its own above-

➊ In a common gravity-fed filter setup, such this one, water is pushed from the pond through progressive densities of filtration media, which house beneficial bacteria. Waste settles to the bottom of the tank, and water returns to the pond via the forces of gravity.

Gravity-Return Filter

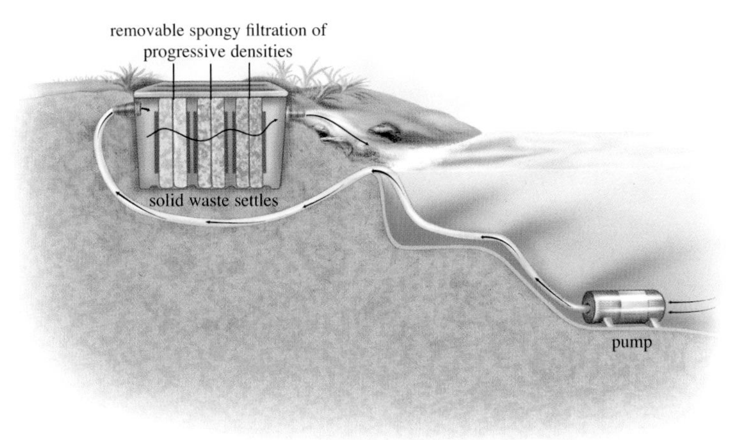

Pressure Filter

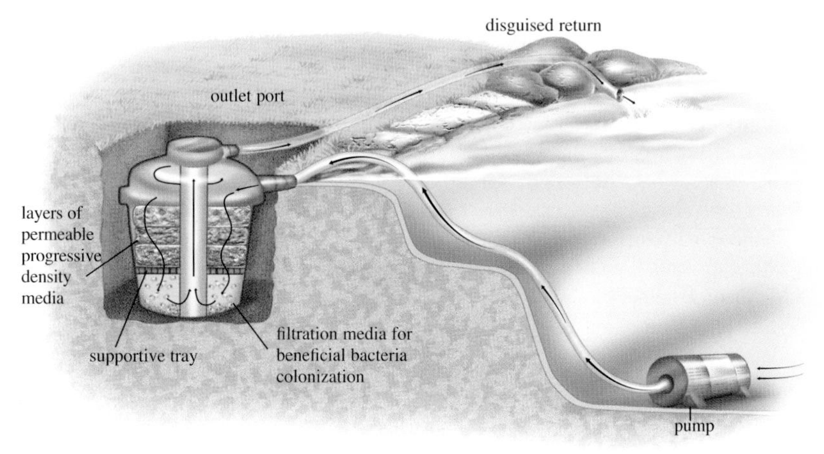

➊ Because pressure filters are designed as sealed systems, the discharge of filtered water can be pushed away from the filtration unit, as shown here.

ground housing. There is no doubt that the multichamber filter is the most efficient of all systems. You can either build your own using concrete or purchase one of the many excellent modular units. The number of chambers will reflect your desire in terms of water health and clarity.

Key aspects of this system are that wide transfer ports are preferable to piping and that the void beneath each chamber is quite large (about 20 percent or more of the chamber). This allows

Gravity-Fed Multichamber Filter

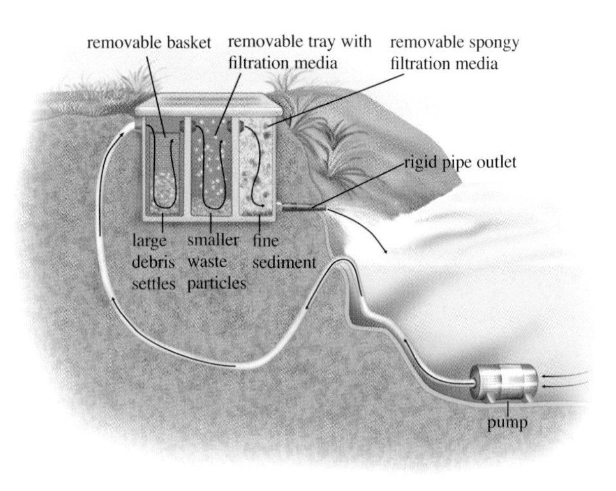

➋ The multiple chambers in this system act as settling tanks for biological material. Each tank helps remove progressively smaller waste particles.

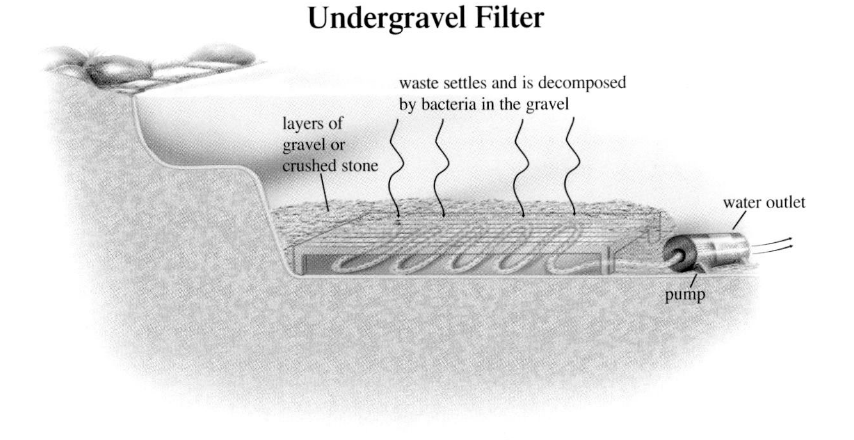

sediment to fall to the bottom where a drain to each chamber (if fitted) allows for removal of the sediment. If no drains are present, the large void beneath the filters is easy to vacuum. The multichamber system is more costly to purchase or build, but its advantage is in its reliability, and performance is worth every cent. Some of these systems have backwash capabilities to further improve removal of detritus.

The *undergravel filter* was once quite popular, but its associated drawbacks have seen its appeal dwindle. When it is working efficiently it is on par with any other good filter. This filter consists of a pipe matrix using pipe of about 1-1.75in (2.5-4.4cm) diameter that lies within a bed of gravel 10-20in (25-50cm) deep on a shelved part of the pond or on the pond floor. In a shelved pond a small wall retains the gravel from falling from the shelf into the pond. A mesh platform supports the bulk of the gravel above the pipe matrix. A submersible pump draws pond water through holes 0.2in (5mm) drilled into both sides of the underside of all the pipes with a space of about 0.75-1in (1.9-2.5cm) between the holes. The pump returns the water to the pond surface via a venturi or other aerator.

The gravel (0.25–0.6in [6–15mm]) develops a large colony of nitrifying bacteria that remove toxins as the pump pulls the water through the gravel. At the same time the gravel traps detritus in a mechanical manner. The size of the filter must be a minimum of 35 percent of that of the pond surface. It is important that the filter matrix fits snugly to the filter wall to minimize the potential for the water to bypass the matrix.

Undergravel filters, though used infrequently because of their cumbersome maintenance requirements, turn the entire bottom of the pond into one giant biological filter.

The negative of this system is that it is labor intensive to keep clean. Once the detritus turns into sludge, the only realistic way to clean it is to remove all the gravel and wash it in pond water. This is a real chore. Raking the gravel (or even using a vacuum) is only successful in the short term because this merely moves it to another location and causes another area where unhealthy anaerobic bacteria can develop. Eventually, a complete clean down is needed.

Filter Size

In the past it was conventional to advise owners that the filter surface area should be about 35 percent or more of that of the pond. Thus a 1,000ft^2 surface area pond needed a filter surface of about 350ft^2. The problem with this concept is that, other than with undergravel filters, it takes no account of the actual surface area of the filter medium. For example, plastic coils have a specific surface area (SSA) that is far less than, say, foam, which in turn has less than that of ceramic rings.

Clearly, if you use a medium that has a high SSA you will not need as much of it as compared with one having a smaller surface area. This being the case, a filter system using high SSA media will be more compact than one using a low SSA media. Ceramic rings will provide about eight times the surface area of plastic coils on a size-for-size comparison. Most commercial biofilters use brushes, foam, and media specially developed by their company.

One of the latest mediums is Alfagrog, developed in Britain and sold in the United States under the name Supra. This ceramic and very porous cinder-type material has a surface area 40-100 times greater than plastic tubing (depending on cinder size) and is already in use by at least one leading American biofilter manufacturer.

Keep in mind that, when shopping for a filter system, you need to know the volume of your pond, rather than its surface area. Your dealer can then show you the various systems that will cope with this. For instance, a 3,000-gallon garden pond will require a much larger and more sophisticated filtration system than will a 300-gallon setup.

Furthermore, all filters have a recommended maximum flow rate that they must not exceed. If the flow rate of the water from the pump into the filtration chambers exceeds the filter's capacity, the filter will overflow. If the pump is too weak and the rate of water flowing into the filter is too slow, the filter will underperform. This means that the flow rate of the filter must also be a factor in choosing the appropriate pump for your garden pond.

Filtration is a very complex subject, and space limitation only allows us to cover it in its basic form in this book. If you plan to have an extensive pond setup, we strongly advise you to seek out a specialist to go over your plans. By so doing, you will be sure the system you use will indeed do its job with capacity to spare.

SKIMMERS AND ALGAE REMOVERS

To complete the subject of filtration, it is useful for you to know about skimmers and algae removers.

Skimmers

Skimmers are useful particularly when a pond is subject to falling leaves. The skimmer is a chamber that contains a removable basket. A pump draws water, together with leaves and other detritus, from the pond surface and passes them into the basket for easy removal. Not all leaves will reach the skimmer as some will sink, but most do, and it saves you continually skimming the pond with a net.

It is best to site a skimmer at the end opposite the one where regular prevailing winds blow as this helps to channel debris towards the skimmer. Keep in mind that a skimmer can trap small fish and other pond inhabitants so it is wise to check it each day.

Removing Excess Algae

There are three methods for removing excess algae; UV sterilizer, magnetic appliances, and ozonizers.

UV light clarifiers (sterilizers) are now a very popular means of removing excess algae. A UV sterilizer comprises a plastic encased bulb of 4–50 watts that connects to the filter system (usually

Skimmers, which draw water from the surface of the pond, aid in removing leaves that have fallen from surrounding trees and shrubs.

➔ Many skimmer units include creative, faux stone covers used to disguise their presence and allow them to blend seamlessly into the design of a pond.

before the biofilter) or is actually located in the biofilter chamber. As water passes through the unit, the UV light kills the algae and causes it to coagulate. In the process, the light also kills some of the unwanted bacteria without causing any undue damage to beneficial bacteria. It is prudent to replace the bulb of a UV light unit (and its O rings and seals) about every six months; otherwise it will cease to function at its best.

Magnetic appliances work to remove filamentous algae that cling to the walls and base of a pond. The magnet alters the carbon ions in the water, and as these ions must be present for algae to cling to the wall, they are thus unable to achieve this. There are two types of magnets, one in which the water passes through the unit and the other, an electronic device, that clamps to the water pipe outside of the pond. However, the water must have a pH of 7.5 or under for success.

Ozonizers serve the same purpose as UV sterilizers, but they are not popular because they subject the pond water to ozone (O_3), which can be toxic. Because the amount of ozone needs careful regulation to make sure it does not harm the fish and plants in the pond, its use is restricted to specialists.

PUMPS

The pump is the very heart of the pond: it's the water's driving force as the heart is the driving force of blood. Without the correct pump for the job to be done, you will find that your beautiful pond will soon look bedraggled. Your filter system will not run efficiently, your fountain will be a sorry-looking dribble, and your waterfall will fall to a trickle. You must choose your pump with great care. A quality pump will perform better, will be more reliable, and will last longer; its running cost will ultimately be lower than that of a less expensive, but inferior, model. Your choice will also depend on where you want to site the pump, what features you value most, the size needed for your particular pond, and head height.

Pump Types and Features

There are two basic types of pumps: submersible and external. Which one you choose will depend on several factors. For the average pond—one with a volume of about 1,500 gallons or less—the submersible is the most popular. It is easy to install, is quiet in operation, and needs no primer to start it. It will cope with most pond features, including simple fountains and small waterfalls. Yet a submersible pump will probably not last as long as an external one and will cost more to purchase and run relative to its performance. For very large ponds with multichamber filters and large water-guzzling features, such as wide waterfalls, the more powerful external pump is the better option. It has the further advantage of being easier to work on than a submersible one (being more accessible). Yet it is noisier in operation, and you have the additional task of building housing for it. (This should be a dry, well-ventilated structure situated as close to the pond as possible.)

In terms of driving mechanism, pumps are either magnetic or oil-filled. A magnetic pump is more energy efficient and environmentally friendly (having no oil to leak). Large external pumps, however, are usually oil-filled to lubricate and cool their moving parts, increasing their ability to operate continuously. If you decide to use a submersible pump within a multichamber filter system, it is best to place it in the last chamber so it pulls the dirty water with suspended particles through the filtration system. If an external pump is your choice, site the pump before the filter chambers so it pushes the water through the system and prevents heavy solids from entering the pump and damaging the impeller.

A number of pumps have a solids-handling capability: all solid material of a given size (which varies by pump) passes through the pump and into the filter system. In such a system, it is best to have a pump that has no prefilter, especially if this is a submersible unit. Prefilters tend to clog easily and therefore must be cleaned regularly to prevent any reduction in flow rate.

When featuring an external pump, do install isolating valves on either side of it so you can turn off the water flow whenever you need to service, repair, or replace the pump. There are two types of isolating valves to choose from—ball and gate valves. Each one functions differently. The type of valve you install will depend on whether you want to use them to simply cut water to the pump or as a flow regulator as well. A ball valve enables rapid water turnoff but is less practical as a flow regulator. A gate valve is excellent for flow regulation but slower if required as a rapid on/off valve.

You can purchase pumps that will operate either below (submersible) or above (external) the water surface, so the choices open to you are indeed considerable. The key features to compare, other than those discussed, are their gallons per hour (GPH) output at variable head heights (we will discuss this shortly) and their energy consumption.

➲ External pumps, such as the one shown here, drive a pond's circulatory system from outside and typically use flexible tubing, which must be disguised.

With the mounting cost of electricity, the amount of energy pumps consume is of concern to many garden pond keepers. Powerful pumps needed to operate large features or extensive filtration units can cost hundreds of dollars a month to operate. You may spend more on an energy efficient pump up front, but, over the long run, it will pay for itself in the lower number of kilowatt hours it uses. Some hobbyists opt to use several pumps in a series to lower the overall electrical draw on the equipment. This conserves energy and extends the life of the pump components.

The pump that fuels your garden pond's circulatory system must run continually throughout the year. Sump pumps, used to pump water to and from the home, are less costly but are not designed for 24-hour use. They work on an intermittent basis in a domestic situation. These pumps quickly burn up when used continually in a garden pond environment. It's worth the extra money to obtain a pump designed for pond use—one that can withstand the rigors of year-round, everyday use.

A variety of materials are used to make pumps. The top-of-the-line models will feature bronze, brass, or stainless steel. You can also buy cast iron or aluminum pumps, which will have lower price tags but will also be less durable. Below these are the small, low-cost pumps that use plastic for many of their components, such as the casing and the impeller that drives the water through the pump.

Do bear in mind when you ponder your pump choice that you can install more than one. Many hobbyists prefer to have a separate pump for waterfalls and fountains rather than splitting and adjusting the output of one powerful pump to cater for the quite different needs of running these features as well as circulation and filtration. Having separate pumps will allow you to switch off fountains and waterfalls at night and during the day when you are not at home. You will also be able to place the auxiliary pump nearer to such features and thereby reduce pipe length. All of this helps keep costs down.

Always discuss your final thoughts with a pond expert, so he or she can give you a balanced opinion based on your specific setup.

Pump Capacity

You also need to consider the pump capacity needed by your particular setup, that is, what your pump output flow must be.

There are many factors that influence output flow. You must begin by calculating, within reasonable parameters, the volume of water the pump must push around your system. Always have a pump whose flow is more than sufficient to do the job; don't get one that is continually working at its limits. You can always restrict the output of a pump, but you can never increase it. (Some pumps have a built-in mechanism for regulating flow; if not, you can place a gate valve on the *outflow* pipe of the pump to regulate it.)

Minute obstacles, such as accumulated debris on the insides of pipes and twists and turns in the pond's plumbing, restrict the water flow and slow down a pump's performance. The total effect of these variations means that by the time the pump has pushed the water through all of the pond's components (filter, waterfall, fountain, and pipes), it ends up returning to the pond at a much slower rate than it was traveling at the pump's immediate outlet.

You need a pump that has the capacity to circulate the basic water volume of the pond (including that of a waterfall system if you plan to have one) about once every hour. To determine capacity, you must take into consideration all features that will impede the water flow. The more important of these are:

- Loss of flow through any filter system.
- Loss of flow due to friction from bends and elbows in pipes.
- Loss of flow due to pipe diameter and length.
- Loss of flow caused when lifting water (its head height) to the top of a waterfall.
- The loss of flow when supplying a fountain.

In addition to the loss of flow factors, you need to consider the stocking level of any fish in the pond. This also influences the water flow requirement. The water's oxygen content must meet the needs of the fish during both cool and hot weather. Remember that as temperatures go up, oxygen levels go down and fish activity goes up, so the pond water flow needs to be quicker at such a time to increase oxygenation.

Calculating Required Pump Size

Bringing all factors together, we can break the pump-size flow requirements down into three components—the flow needed to circulate the pond's total volume (a) through the filter system and back to the pond, (b) through any fountains, and (c) into the waterfalls, assuming the latter two features are part of the pond design.

Pond pumps are rated for their ability to cycle or "turnover" a certain volume of pond water in a matter of time. For example, a pump that claims 3,500 GPH means that the pump is capable (under ideal circumstances) of pumping 3,500 gallons of water per hour. This is why it is vital to determine the total pond volume before purchasing a pump. The more components and features a pump must support, the higher the rating of the pump. You can plan ahead for this by adding to the known volume of your pond to accommodate the added power necessary to operate the system.

For the pond and a mechanical filter with no fish and plants, the rate you can work on is a turnover of total water volume of once per hour. If the pond contains a few medium-size fish as well as a few plants and a biological filter, increase the pond's total water volume figure amount by 50 percent. If the fish are koi, increase the turnover rate by at least 200 percent. If your pond is in direct sunlight for about half of the day, add 10 percent to the total pond volume. If it receives direct sunshine throughout the day, add 20 percent to the total pond volume.

There are many variations in pond fountains, and you can work on the premise that these will require a pump flow of 50-300 gph (gals are US, multiply by 0.833 for UK gals) for most small to medium sprays or statuaries.

Waterfalls are a work of art in themselves, and the variables are legion. The factors to consider are the width of the waterfall spill (weir), the depth of the water you want to pass over this, and the height of the waterfall (its head) in relation to the pond surface. Every inch of spill width will require about 100 gallons at a depth of about 0.25in (6mm).

This increases dramatically the more water depth you wish to have so that at 0.5in (12mm) the flow need is about 85 percent more water; thus the 100 gallons becomes 185 gallons! If you wish to have a 12in (30.4cm) weir at a depth of 0.25in (6mm), this will require a pump capacity of 1,200 gph. If there are two or more spills en route back to the pond, then the 1,200 gph will repeat for each spill so that with three spills a pump flow of 3,600 gph is necessary.

Pump Head Height

The term *head* applies to the vertical distance a pump will push water up a pipe. Each pump has fixed limitations on this at variable heights. It has a maximum capability after which it is unable to lift the water any higher. At its maximum head, the water flow is considerably less than the pump's actual nonlift flow. Each manufacturer stipulates the flow rate at certain head heights, and each indicates the power consumption in watts. See Table 3, which gives an example of power consumption for a well-known brand pump model.

TABLE 3 POWER CONSUMPTION IN WATTS

Max. Flow	Flow at 1.65ft	Flow at 3ft	Flow at 4.95ft	Flow at 6.6ft	Flow at 9.9ft	Watts Rating
1,584	1,505	1,409	1,306	1,188	950	100

There is now one more issue to consider with regard to head height. It relates to the length of pipe between the pump and the outlet pipe to the waterfall. This length is the distance from the pond surface to the waterfall outlet. Every 10ft (3m) of piping equates to 1ft (.03m) of head. If the waterfall head is 4ft above the pond surface and there are 25ft (7.6m) of piping to reach each, then the total head is 6.5ft (2m). Each pump will stipulate the size of pipe it requires for the pipe to handle the pump's outflow.

In general, the larger the pipe diameter the better because it reduces pipe friction loss as well as pressure on the pump. When the run of piping exceeds about 15ft (4.5m), it is advisable to use the next size higher in pipe diameter. If this is greater than the outlet/inlet of the pump, a reducer will enable the larger pipe to fit the pump connection so that it's compatible with the pump volute (see Table 4).

TABLE 4 PIPE WATER FLOW (GPH IN US GALS) AT VARIOUS DIAMETERS

Pipe Diameter	Gallons
0.5in. (12.5mm)	250
0.75in. (19mm)	500
1in. (25mm)	750
1.25in. (32mm)	1,375
1.5in. (37.5mm)	1,875
2in. (50mm)	3,125
2.5in. (75mm)	4,375
3in. (75mm)	6,875

Note: Internal smoothness of pipes, pipe length, elbows, bends, inclination in the pipe level, and pump type and efficiency will all influence flow. The figures stated above are merely guideline approximations.

When you have completed your calculations, it's time to select your pump. You are not likely to find a pump that matches your needs exactly. So choose the next size above that which meets your water needs. That way, you have the little extra leeway we have discussed. When comparing pumps, you will find the power consumption in watts does vary between different companies and models, even at the same maximum flow rate. It is therefore prudent to consider not only the energy cost but also the pump warranty, anticipated longevity, and total features. There is invariably a subjective trade-off just as there is when comparing cars or other items we wish to purchase.

Pond Volume

You need to know the volume of a potential pond, so you can purchase a pump of appropriate power to move the water through the filter system—assuming you plan to install a filter, which is strongly recommended. Knowing the volume is also necessary if you ever need to add medicaments or other treatments to the pond. To calculate the volume, multiply length by width by depth.

For example, if a pond is 15 x 10 x 4ft (4.6 x 3 x 1.2m), its volume is 6,300ft (16.63m).

A cubic foot holds 7.48 US gallons of water so the sample pond holds 600 x 7.48 = 4,488 gals. This equates to 17,000 liters, or 3,738 UK gals. See the Appendix for useful conversions, so you can calculate any volume in either metric or nonmetric units, whichever you are more comfortable using or need.

As stated earlier, the depth of a pond is important because it affects the water temperature. A deep body of water—more than 4ft (1.2m)—remains stabler than a shallow one. It is also better for the fish psychologically and physically, allowing them room to exercise. During the winter period, a deep pond offers the fish a haven in its lower levels when the higher levels are frozen.

CHAPTER 5

Waterfalls, Fountains, and Watercourses

The sights and sounds of waterfalls, fountains, and watercourses intoxicate the senses and soothe your weary soul. Waterfalls and fountains not only create visual appeal and a relaxing environment but also aid in aeration of the water, making for a healthier environment for fish, plants, and other backyard wildlife. Adding one to your garden pond may not be as difficult as you think. With some careful planning, and creativity, the possibilities are endless.

The addition of a waterfall to your garden pond brings relaxation to the soul and beneficial oxygen to the water.

WATERFALLS

Whether added to a formal or informal pond, there are a few guidelines to keep in mind when adding waterfalls. First and foremost, make sure that the size of the waterfall corresponds to the size of the pool or basin into which it will be flowing. Avoid installing a tiny trickle that empties into a vast open pond, and certainly avoid a torrential cascade into a diminutive puddle. Either of these mistakes will result in unnatural ponds you will undoubtedly be unhappy with.

There are no hard, fast rules for the size and height of a waterfall. It is largely a matter of personal taste and involves a certain amount of trial and error. With common sense and creativity in mind, create something you can live with. You may want to build Victoria Falls in your backyard, but do keep in mind certain considerations, such as the fact that high, raging falls that empty into shallow pools will speed evaporation and churn up your pond water, turning the immediate area into a chaotic whirlpool instead of a relaxing backyard retreat. Keeping falls low and allowing them to empty into deep pools creates a good balance.

Second, try to imagine the finished, mature pond, complete with plantings and pond fish. Be sure that the scale of your waterfall will still allow for calm areas where fish can seek shelter and water lilies and other plants can grow undisturbed.

There are a variety of ways to build a waterfall. As with liners, waterfalls can come in preformed varieties, which can be installed at any pitch for the desired effect. Placed in the same way as their larger counterparts, these waterfalls simplify the process and offer instant gratification. Flexible liner and rocks of varying size can also be used to create virtually any configuration in an informal setting. Their installation requires an artistic eye and a flare for creativity. For more formal features a variety of materials can be used, from brick and tile to acrylic and industrial materials.

Whether features are constructed with flexible liner and rocks, fiberglass, or other materials, make sure there is proper alignment of the biological filter output or recirculating plumbing from a pump component before the installation gets underway.

Free-form, naturalistic waterfalls can be the most challenging to construct, that is, if your goal is for the feature to appear as natural as possible. Using nature as a guide, choose an incline that fits the contours of the existing space. When building up the incline, keep in mind any areas where large rocks or boulders might be placed, and create spaces to receive them. When large boulders or stones are placed near each other in a waterfall situation, they create a gushing effect. In addition, consider any flat spaces that will receive flagstone, if you are planning on creating a sheeting waterfall. Allow pools for the collection of water that will fall from a dramatic height, allowing it time to collect before running over the next cascade.

Pack the excavated area well to avoid later settling of the liner material. Be sure to fasten the edges of the liner and secure it to the water source to prevent leaking of the feature and subsequent draining of the pond.

Waterfalls that appear as they would in nature often require larger boulders, such as the one shown here, and take the most careful planning and rock selection.

Dry fit all the stone you will be using to naturalize the feature, making sure that the waterfall will look aesthetically pleasing and create the desired effect when completed. When you're satisfied with the construction, you can anchor the rocks with expanding urethane foam spray and waterproof sealant. This will ensure there are no leaks and prevent the water from undermining the structure of the falls.

If you are using expanding "rock foam" (as it's sometimes marketed), be sure not to use too much. It expands over the course of several hours and can ooze out from between cracks, requiring you to go back with a sharp knife or saw to trim the excess away. While the foam is still tacky, sprinkle sand or fine gravel over it. It will adhere to the foam and cloak the finished mortar joints, creating an even more natural appearance.

PRINCIPLES FOR ACHIEVING A NATURAL LOOK

If you're looking to create a waterfall as it would appear in nature, here are a few principles that will help you get there. Nature makes it look easy, but thousands of years of water trickling over stone is not as effortless to achieve on your own.

Select the right type of stone.

Use stones referred to as river rocks, river flats, and river rounds that are smooth or worn with the passage of time and water.

Avoid rocks that have never seen water.

Avoid sharp angled edges, which are often found on quarried granite from the insides of mountains; such rock will never look as though it belongs in a naturalistic waterfall.

Avoid disjointed placement.

Rocks should be placed in relation to one another, either perpendicular or parallel. Skewed angles look unnatural and disjointed.

If you are using a product other than natural stone and the desired effect is meant to be more contemporary, the construction phase can really be an adventure. You may invent your own techniques as you go along, creating a truly one-of-a-kind appearance.

A word of caution: be sure to dry fit every piece of material before you decide on its permanent location. You may even want to set all of the materials in place and then dump a bucket of water or turn on the garden hose to see how each piece position you've chosen will react to the flow of water. You may want to conduct several trial runs like this to make sure you're absolutely satisfied with the effect you've created before you anchor your waterfall into place. This will avoid waste of materials and time and ensure you will enjoy your masterpiece for years to come.

FOUNTAINS

Nothing accomplishes the sound of moving water better than a fountain. As droplets of spray catch the sun by day and moon by night, they bring a magic to the garden pond that is unparalleled. The accents offer not only beauty, but function as well. The action of moving water introduces dissolved oxygen to the water, benefiting the plants and fish that call your pond home.

The many options of fountains available on the market today, which include fountain heads and sculptural fountains with waterspouts built in, all function with a simple submersible pump. Consider a few things before making your decision.

If your pond is located in a windy area, you'll want to think carefully about the pattern and height your fountain will produce. Too high or too wide of a waterspout and our pond will soon be drained as the wind blows it out of bounds, soaking surrounding landscaping and furniture.

Too fine of a water spray in a sunny location will also cause evaporation and create a battle to keep the pond level consistent. A good rule of thumb is to make your fountain's waterspout height no more than half the width or diameter of the pond.

Another consideration, as with waterfalls, is to be sure not to create too much water disturbance that will prevent you from growing any water plants you may be planning to add. Be sure there are sufficient areas of the still water many of these plants need to survive.

Statuary and wall-mounted fountains, also known as spitters, add an element of sophistication and whimsy to garden ponds. They are typically installed in the same way as traditional fountains and often retail in kit form to make them manageable for the do-it-yourselfer.

Many commercially available fountain kits come included with pumps that have adjustable flow valves for adjustable spray heights, as well as interchangeable fountain heads that create different patterns. Fountain heads need regular cleaning to

➊ Fountains take many forms, such as this cascading terra-cotta piece, which introduces vital oxygen as it returns water to the pond.

➔ Easy-to-install spitters add an element of fun and function to a garden pond.

prevent clogging, which spoils the intended effect and can burn out pumps if left stopped up for extended periods of time.

Installing a simple spray fountain is typically the easiest fountain option available to consumers. Since it consists of a small pump, preplumbed pipe, and threaded fountain head, you just pick the location and lower it into the pond. You may need to create a supportive pier of bricks on the pond bottom if you have a deeper pond or want the fountain head to rise up out of the water. These fountains are often available as floating units, allowing you to simply screw on the appropriate fountain heads and float them into the desired location.

Statuary fountains and spitters can be a bit more complex, as they usually require a supportive structure. Whimsical figurines usually come predrilled and ready to receive flexible tubing. Select the proper gauge and insert the tubing into the hole; seal the tubing in place if it does not fit tightly. Attach the other end to the output of a small submersible fountain pump. Check for leaks after the pump has been allowed to run for several minutes. When you're sure there is an adequate seal between the fountain device and the pump, either mount the fixture on a brick pier out in the pond or fasten it to the pond margin or nearby wall. Disguise any visible tubing, and you're in business.

WATERCOURSES

Watercourses can add much depth and character to your garden pond. Often, in flat spaces without much contour, waterfalls and fountains look contrived. Yet in these same situations, streams look right at home. They accomplish the same beneficial effects as waterfalls and fountains, adding a treat for the senses and an oxygen-rich environment for better water quality.

Only slight elevations are needed to keep streams flowing—no less than 2 inches per 10-foot run. The elevation should change just gradually enough so that water still pools in the streambed after the pump has been turned off. This prevents the streambed's mortar joints or liner from drying and cracking in the sun.

Many preformed watercourses are available on the market and can be installed in the same way as other preformed garden pond accessories. But with a roll of flexible liner or bricks and mortar and a little imagination, the sky's the limit.

Whether you desire an informal winding stream or a formal well-defined structure, there are several construction techniques that change the overall effect and feeling of a watercourse. To cause a stream to run faster, make the banks closer together. To slow it down, move them farther apart. To create gurgling rapids, add stones and boulders. To accommodate bog plants and other submersed plants, create a wide stream with shallow areas and shelves near the banks. Care should be taken, however, to avoid making a watercourse too wide, as stagnant areas of water will be created, leading to algae growth and mosquito incubation sites. See Table 5.

As with waterfalls, which share similar construction techniques, it is wise to dry fit and test water flow several times during the installation phase. Try to mimic the amount of water that will be flowing over the surface, to the best of your ability. This trial-run phase may cause you to repeat the construction process of the entire watercourse several times over to achieve the desired effect. It is well worth the time and consideration and will prevent you from having to disassemble and rebuild the entire thing down the road. When you anchor or seal the materials that make up your watercourse, it is likely that they will stay in that position for years to come.

TABLE 5 EXTREME SPECIAL EFFECTS

Desired Effect	Design Required
Faster Watercourse	Move banks closer together.
Slower Watercourse	Move banks farther apart.
Gurgling Watercourse	Add stones and boulders.
Planted Watercourse	Move banks far apart and provide shallow areas for plantings.

CHAPTER 6

Electricity and Lighting

The availability of electricity is clearly a must for a pond in which filters, fountains, lights, a vacuum, and maybe a heater and other accessories are part of your operational and landscaping plans. There is no doubting that the inclusion of lighting both in and around a garden pond will turn your daytime mini-paradise into an enchanting and totally different-looking vista once it is dark.

Cascading waterfalls and fountains sparkle with reflective colors in total contrast to their daytime appearance. Soft lighting gives the pond itself a totally different attraction as the fish meander in and out of the light-and-dark contrast in their watery world. Paths and foliage subtly toned under white, orange, blue, or green lights allow you to enter an almost mythical world far removed from the hub and stress of day-to-day life that is the norm for most of us.

ELECTRICITY

Ensuring an adequate electrical supply to a garden pond can be a hair-raising experience for the novice. Considering the maze of municipal codes and potential hazards involved with installing this part of your backyard oasis, hiring a professional electrical contractor may be the best decision you ever make.

If you're going it alone with electrical installation, be prepared with sound information that will keep you safe and also keep your work within the letter of the law. Take some time to familiarize yourself with your municipality's electrical codes and any federal regulations that may dictate how outdoor electricity is to be handled. Special care needs to be taken when installing outdoor electricity, particularly when working near water. If you're not 100 percent confident in your abilities, hire a professional.

Before wiring permanent electricity to your garden pond's various components, the entire pond should be complete, with all devices installed in the locations you have chosen to put them. You can use an extension cord from the house or garage to be sure each of the devices is functioning properly, but it is essential that the setup be nearly final before the first cable is laid.

This precaution will prevent mishaps, such as running into a tree root or boulder, which would force you to move a component

Submersible lights, such as the ones installed here, extend the enjoyment of a garden pond past sundown and create an atmosphere for relaxing or entertaining.

to another place in the garden pond, putting your previously established power source out of reach.

All exterior electrical sources should be equipped with a ground fault circuit interrupter (GFCI), which automatically shuts off the electrical supply to your pond if fluctuations in the electrical current or moisture are detected. This safeguard prevents your electrical devices from overloading and burning up and keeps you and your pond's fish from potentially deadly electrocutions.

You need to calculate your pond's overall energy needs to make sure you install the hardware to handle the electrical draw that will be required to operate all the components. The electrical devices used in your pond should be labeled as drawing a certain number of amps or watts. Compare the device's power requirements with the number of amps or watts your GFCI can handle. A standard, 20-amp GFCI can handle up to 2,400 watts.

THE MARK OF CONSUMER SAFETY

Underwriter Laboratories is the most widely known safety certification organization in the world. Nearly every electrical device and appliance on the market bears the UL seal of approval. Each product with the UL seal has been tested under more than 800 safety conditions and certified safe for consumer use, when used as directed by the manufacturer. Electrical components not marked with this seal or the seal of another independent safety inspection laboratory may mean added risk, even when used to the manufacturer's specifications. Look for some sort of safety seal, either stamped into the device itself or written on a tag attached to the cord of the device.

You can determine how many watts your components will require with the following formula: amps x volts = watts. Add the watts for each of your components together. If the total number of watts exceeds 2,400, you must install a higher-rated GFCI. Failure to do so will result in your circuit overloading and tripping every time the system is switched on.

For maximum protection and safety, electrical lines from the main source to your pond's power outlet should be run through PVC conduit or metal tubing and be buried at least 18 inches beneath the ground. This step will protect electrical cables from being damaged by motors, spades, and other gardening implements.

Place the power box with outlets nearest to your primary electrical device—in most cases, your pump or skimmer box. Be sure to choose a deep-style box that will accommodate the GFCI and all electrical cords and plugs. The box will need to form a complete seal when closed, with every device plugged in, to avoid moisture seepage into the box, which will cause faulty tripping of the circuit.

The box and any lighting transformers can be mounted a few inches off the ground on a low post. Be sure to put them in a location where you can easily camouflage them with boulders, plants, or other accessories, leaving you with an aesthetically pleasing view of your garden pond. Consider installing an indoor switch that will enable you to operate the power source from a remote location.

LIGHTING

Made possible by electricity, lighting brings an entirely different dimension to a garden pond, allowing busy homeowners to enjoy a backyard oasis during the only free time they may have—nighttime. There are several different types of lighting available for the garden pond environment, specifically made for use near or in the water.

Submersible lighting can be angled to feature underwater plants or swimming pond fish. Placed behind a waterfall or at the base of a fountain, lights create sparkling ambiance. Surface

Lighting can change the entire look of a garden pond, offering a vantage point on the underwater world not possible during the day.

lighting showcases plants whose textures are otherwise lost during the night hours. Landscape lighting creates romance and drama, and all options can be changed with color filters to offer an unlimited number of combinations.

Low-voltage lighting systems run on 12 volts and require a transformer, connected to your main power source, which reduces the standard 120 volts to the proper amount to run the lighting. Low-voltage systems are often sold in do-it-yourself kits and are widely available for pathway and landscape applications. These

Directing lights toward architectural details such this bridge highlights its graceful shape and illuminates the structure for added safety.

systems typically include the transformer, the cable, and several bulb assemblies. Simply hook up the transformer, run the cable, decide where you want to place the lights, and attach them at the proper location along the cable. Their smaller size means they can be tucked almost anywhere. These fixtures use automotive-type bulbs, and their locations can first be tested using a bright, waterproof flashlight.

Standard 120-volt lighting units, plugged directly into the main power supply, are often much brighter. These compact high-intensity lighting fixtures use primarily halogen bulbs and are commonly installed by garden pond enthusiasts. These lights can easily be tucked among rocks or behind plants to obscure their view during daylight hours.

Larger floodlights can also be used above and below the water level to illuminate large pools or vast areas of the garden pond. These lights typically have large housings to accommodate their powerful bulbs. If they are used above the water, disguising them can be a bit of a challenge. When tucking them behind dense landscaping is not an option, you can create recessed light vaults below the ground, with containers to receive the units. Be sure to add drainage holes so the vaults don't fill with water after heavy rains. Include metal grates as well to keep the vaults from filling with debris. Be aware that these lights may also create heat, which can raise water temperature and burn nearby delicate plant foliage.

THE WAVE OF THE FUTURE

Advances in technology have made even more options available for shedding light on garden ponds. Many pond supply companies now offer submersible fiber optic lighting systems. Fiber optic systems create light from one bright light source connected to the main power source outside the pond environment. From that source, light travels down a glass or plastic fiber cable through a series of internal reflections. Light emerges from the end of the cable as a bright point, which can be intensified by a small fixture. Light sources contain several cables, meaning several points of bright light for every one lightbulb operating in a remote location.

The many benefits of this form of lighting come from the fact that no heat or electricity is being introduced into the water, and only one lightbulb needs to be replaced, which can be done without reaching or wading into the pond. The compact fixtures at the ends of the flexible cables can be placed in very tight spaces where other fixtures will not fit. This high-tech option is more expensive than traditional methods of lighting but worth it if you're willing to spend the funds.

Light emitting diode (LED) technology is yet another option for underwater lighting. Though currently much more expensive than existing lighting technologies, LED offers an exciting new option for garden pond enthusiasts. Originally developed for use as automobile indicator lights, LED technology is quickly making its way into interior and exterior applications. LEDs convert electricity directly into light by passing it through a solid-state semiconductor. The resulting glow produces no heat and uses a tiny fraction of the electricity used for other lighting options. Waterproof LED lights can be simply screwed into existing traditional light fixtures and are designed for years of continuous use.

CHAPTER 7

Pond Construction

Careful consideration of all the components that go into a garden pond oasis will only get you so far. Eventually, you will have to get to the down-and-dirty process of construction. Turning an ordinary environment into a pond you will enjoy for years to come often requires just plain brute force. Diagrams of the finished project help. Consultations with professionals are invaluable. But a do-it-yourselfer's best ally is a good strong back.

CREATING A BLUEPRINT

The first step is to create a blueprint for your pond, sketching in existing structures and the proposed pond and its components and indicating the various elevations for construction. You will need a grid—a page of graph paper, available at most art supply and office stores, or your own rendering of squares on a piece of plain paper. Each square will be equivalent to 6in (15.2cm) (larger projects may require a scale of 1ft [30.5cm] per square).

Begin by drawing the garden pond space as it appears now, including all surrounding buildings, landscape features, and decking or patio constructions. Be sure to note the locations of existing underground plumbing, electrical, and gas lines to avoid violating municipal codes or accidentally damaging your own utility services—a costly and easily preventable mistake. You can also contact your local utility company, which will arrange to visit your property and mark out existing utility lines with paint or flags.

Next, sketch out the pond area as you intend the finished project to appear. Note the locations of skimmer, filtration, and pump components and the plumbing that will be required to operate these systems. For an informal pond, keep in mind any gradations in elevation you plan to create with excavation and stonework, represented by lines inside the outline of the pond itself. Indicate the depths of each level created by these elevation lines. If the graph is becoming too cluttered, create a separate one for depths.

Making a small-scale version of the surrounding environment will ensure proper placement of the pond and help the construction process proceed smoothly. Often, problems can be worked out on paper before they cost you time and money during the construction phase of the project.

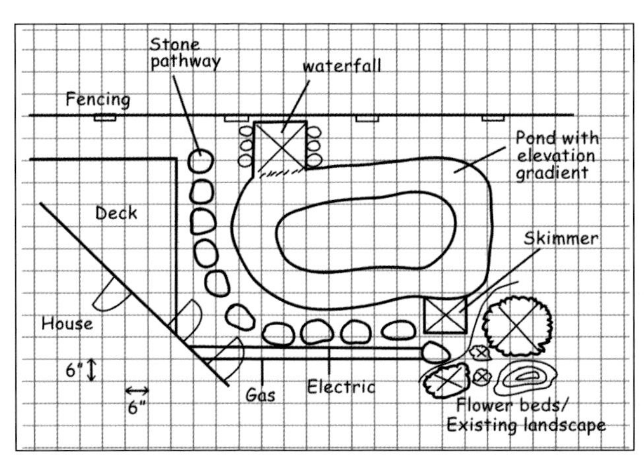

➔ Creating a scaled blueprint of sorts, such as the example shown here, gives you a valuable guide to follow throughout the construction phase. Existing structures should be included to ensure proper placement of your pond.

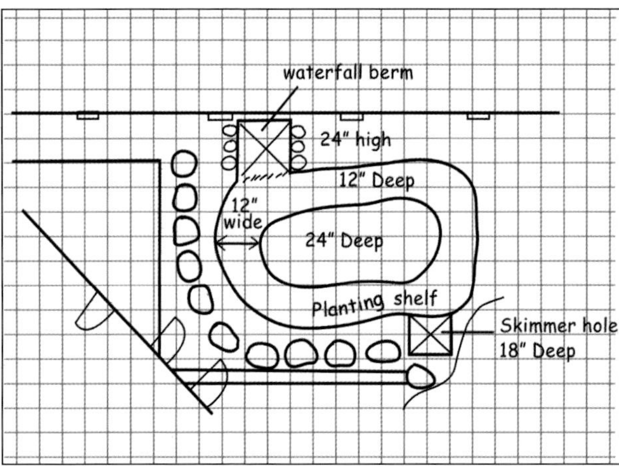

➔ An elevation gradient diagram such as this one will remind you how deep to dig or how high to mound the soil as you excavate to achieve the desired effect.

MARKING THE LAWN

Armed with a blueprint of your garden pond, a few flags or wooden stakes, a mallet, a can of spray paint, and a measuring stick or tape, you are ready for the next phase of construction. Transferring the shape of your garden pond from paper to ground ensures that the finished pond matches the oasis of your dreams.

Plans in hand, set out the perimeter of your garden pond, keeping in mind that the distances need to be as precise as possible. Measure to be certain, from existing structures in the yard as they relate in your blueprint to the intended perimeter of your garden pond. For instance, if you intend for the far edge of your pond to be 1.5ft (45.7cm) from the corner of an existing flower

➔ Using the blueprint as a guide, lay out the shape of the garden pond, with the aid of a measuring tape to perfect the dimensions. Use stakes to mark reference points and soil, lime, or paint to connect the dots, as shown here in the example.

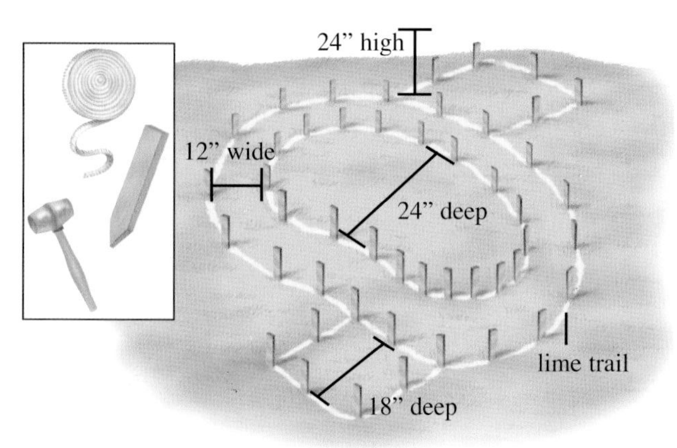

bed, take your measurement with a measuring tape from that flower bed and start there. At several points on the outside edge of the pond border, place wooden stakes or garden flags to mark the distances as they relate to your plans.

Once several of the distance points have been mapped and marked, you will need to connect the dots, so to speak, with a can of brightly colored spray paint. The contours of informal ponds can be "sketched" freehand. You can also use a trail of lime or soil to map out the pond's perimeter for a less permanent option with low environmental impact.

The elevations can then be painted on, with heights noted and allowances made for plumbing, submersible components, and planting ledges. Blowing up your plans on a grand scale gives you guidelines to follow for excavation and construction. It keeps you from digging blindly and making aesthetic mistakes that result in a pond with which you are not satisfied.

If you are installing a preformed pond, you can simply drag the form to its desired location and, while it rests on the ground, transfer the contours and elevations straight down from the sides of the form onto the grass below.

Examine your handiwork to be sure the outline of your garden pond matches your vision of the completed project. It's far easier to correct errors at this early stage than to change your mind after the earth has been removed from the space.

EXCAVATING

Small-scale excavation projects can be completed with a shovel and some elbow grease. For larger projects, you may want to consider hiring a professional excavator or renting heavy equipment for the weekend.

It's best to start at the deepest part of the pond and work out to the shallower portions, building up elevated areas for secondary pools, waterfall basins, and watercourses as you go with the dirt you've removed. The rough shape is carved out in this initial phase.

As mentioned earlier, during the excavation phase of a project, there could be obstacles lurking beneath the surface. Large boulders may throw a wrench into your plans, requiring the use of a jackhammer or perhaps forcing you to relocate the entire project.

Tree roots present their own special set of circumstances. When building a pond near an existing tree, be wary of the tree's drip zone—the area extending from the tree's base to the farthest point where moisture drips off the tree's canopy. The underground area directly beneath a tree's drip zone contains the tree's strongest roots. Cutting any of these roots in order to accommodate a new garden pond could cause distress to the tree, which could, in turn, kill the tree. Older, more established trees are less able to withstand root cuttings than younger trees.

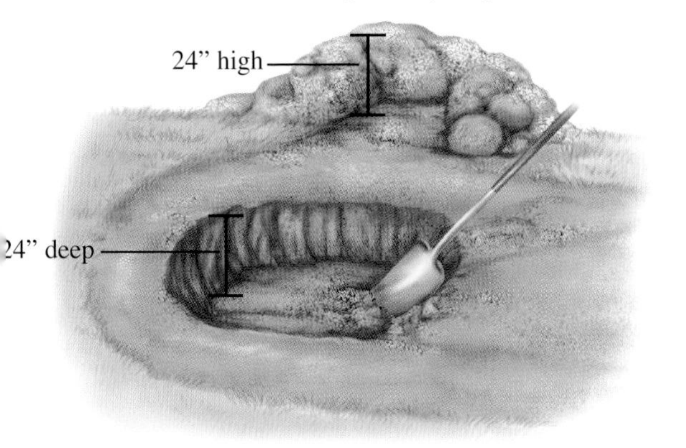

Following the lines and elevation gradient, excavate the soil. As the soil is removed from the deepest points, it can be used to build up any waterfalls or watercourses, as demonstrated by the diagram.

SHAPING AND LEVELING

Once the overall shape and approximate elevations have been roughed out, there is no substitute for the old-fashioned shovel and trowel. Shapes, contours, and planting shelves must be sculpted by hand. Consider whether your garden pond will feature bog plants, marginal plants, or water lilies (discussed in chapter 9), as each of them will require different water depths involving basins and shelves, which must be created at this phase.

Leveling the elevations is also crucial at this point. Each of the planting shelves must be plumbed with a level to ensure stability and adjusted accordingly, using a shovel and tamper. For leveling of planting shelves and pond edges, you'll need to construct a simple device. Insert a wooden peg in the center of the excavation, at the pond's lowest level. Make sure the stick itself is level in its perpendicular angle to the surrounding ground. Hammer it down so that the top of it is at the level of the elevation you are checking for plumb. Set a flat board atop the peg, and rest the other end on the pond's edge. Place a level atop the board. This will allow you to swing the board freely about its "axis," with one

SEVEN EASY STEPS FOR USING A SITE LEVEL

A site level is a device used often in the construction industry. It consists of three main components: the site level, the tripod, and the measuring stick. These devices are used to measure exact elevations and depths during the excavation process. Why does it matter if you use one? For larger projects (half an acre or more) a simple extra inch of depth in your garden pond could result in thousands of extra gallons of water, which you may not have calculated when choosing your pond's electrical components.

1. **Set up the tripod in an area where it will remain undisturbed.**
2. **Check to be sure the tripod itself is somewhat level.**
3. **Attach the site level to the tripod with the large center screw.**
4. **Use the leveling screws on the bottom of the device to make it level in all directions by situating the level bubble within the marks indicated on the top of the site level.**
5. **Bring the crosshairs into focus as you look through the site.**
6. **Look through the site as one person holds the measuring stick at several locations around the perimeter and various elevations, before you begin excavation. Record your measurements carefully on your paper plan. This helps establish the basic topography of the site and sets a baseline for any changes you'll be making.**
7. **During excavation, as the stick is placed in a lower spot, the values increase. As it is placed in a higher elevation, the measurements decrease. Compare your new measurements to be sure the difference from the base line match your desired excavation plans exactly.**

edge always resting on the excavated earth, checking the bubble continuously for variations in height or depth that need to be corrected. If your pond is a large one, with many different levels, you might consider using a site level (see box).

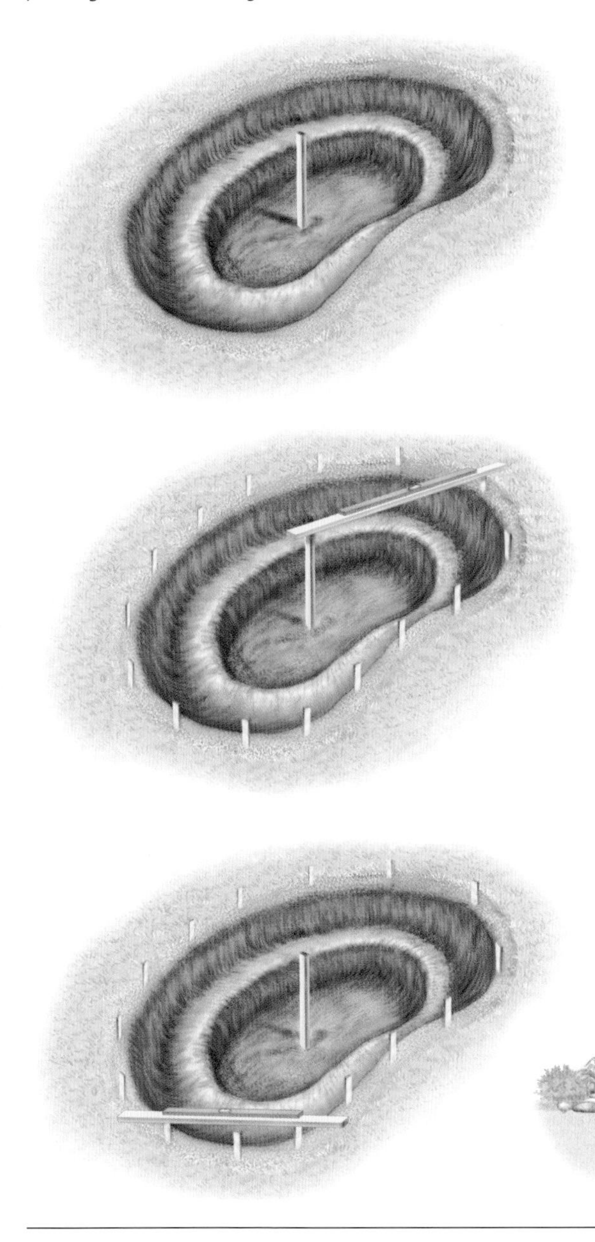

➔ 1: To level the elevations, insert a straight piece of lumber in the center of the excavated space. The top of the stock should match the desired water level.

2: Place another straight piece of lumber atop the center board, and rest it on the edge of adjacent marker pegs.

3: Check all planes for level, and adjust the excavation accordingly.

Below: When making final adjustments to the contours of your garden pond, visualize the finished project and plan for the features you want to incorporate later.

PLACING MECHANICAL/PLUMBING COMPONENTS

Decide where you want the pond's surface to fall and make sure the inside edge of the skimmer intake is at that level, situated in the hole you dug during the excavation phase. Likewise, the level for the filter return should be located to allow for oxygenation of the water as it returns to the pond. If you are using a biological filter falls unit, secure it in the excavated hole you created atop the waterfall berm during the excavation. If you're installing a settling tank filtration system, excavate spaces to receive the tanks and install them. Place pumps in pump houses you have created, if using large external models. The locations of bottom drains and overflows, if applicable, should be finalized and plumbed before the pond is lined. Bury and secure any and all subterranean plumbing connecting your pond's components a minimum of 12in (30.5cm) deep to avoid damage from normal lawn activities. Be sure these mechanical and plumbing devices are secure before proceeding to the liner phase.

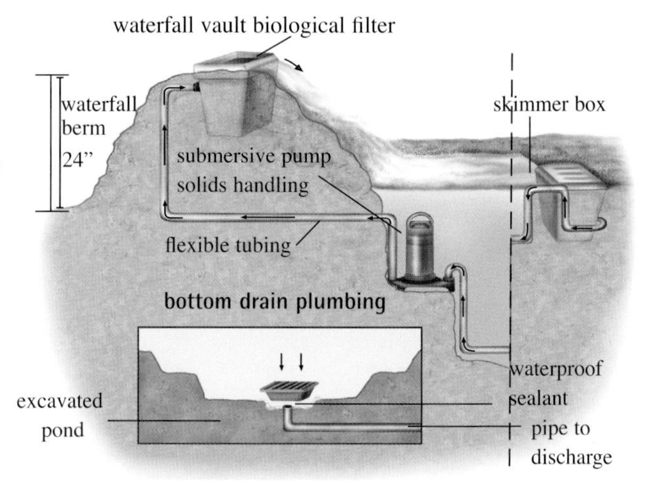

➔ When placing mechanical components, remember: 1) The elevation of the intake will determine the water's surface level, meaning it will have to be buried (as shown here). 2) The height of the waterfall box will determine the pitch of the cascading water, meaning it will have to be received by an elevated berm.

Inset: Be sure to seal all plumbing to create a closed system and seal the bottom drain, if included, as indicated.

REMOVING DEBRIS, PACKING, AND CUSHIONING

After elevations have been finalized, the area can be raked to remove loose rocks, tree roots, and debris that may puncture a flexible or preformed liner. The contours can then be packed firmly using a commercial compactor or tamper. A layer of sand or geotextile underlayment can be added for increased cushioning and protection from sharp objects.

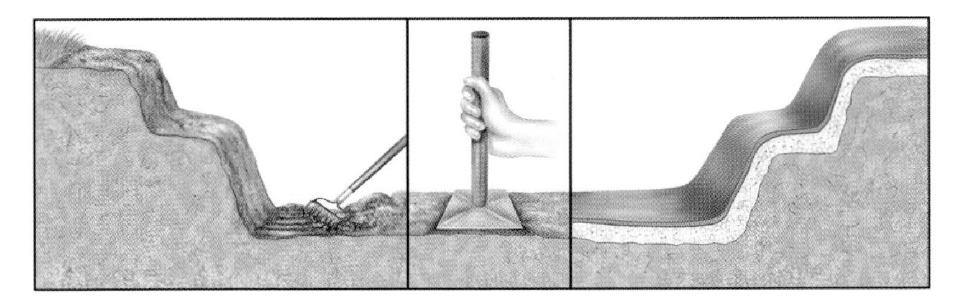

➊ Step 1: Rake any debris and sharp stones from the pond site.

Step 2: Use a tamper to compact the earth and create a smooth surface to receive the liner.

Step 3: Drape geotextile underlayment and flexible liner over the excavated site, easing it into the contours of the pond.

SECURING THE FOUNDATION

With the shape finalized and ground prepared, the space is ready to receive the liner that fits your chosen pond style. Bottom drains and plumbing openings, when present, should be covered to pre-

THE BOTTOM DRAIN ISSUE

Should all garden ponds incorporate bottom drains? This is a currently contentious issue among professional pond builders. Some professionals state, emphatically, that no pond should be without one, while others feel just as strongly that they can only lead to disaster. Consider both sides of the issue carefully before deciding whether to plumb the depths of your own garden pond.

ARGUMENTS FOR	ARGUMENTS AGAINST
Several inches of sludge collect on the bottom of a pond, which then periodically needs to be flushed out, via a bottom drain.	**Sludge collects more slowly in a gravel-bottom pond that provides surface area for beneficial bacteria, which decompose it.**
A bottom drain is the fastest, easiest way to drain a garden pond for its annual cleanout.	**Pumps and siphons can be used to drain a pond. An ecologically balanced pond will not need to be drained totally every year.**
Bottom drains are easy to install and can be used, in conjunction with plumbing, to direct nutrient-rich water to other areas of the garden.	**Siphons and hoses can also direct water from a pond to a nearby garden and do not require an added cut in the flexible liner that could leak at a later date.**

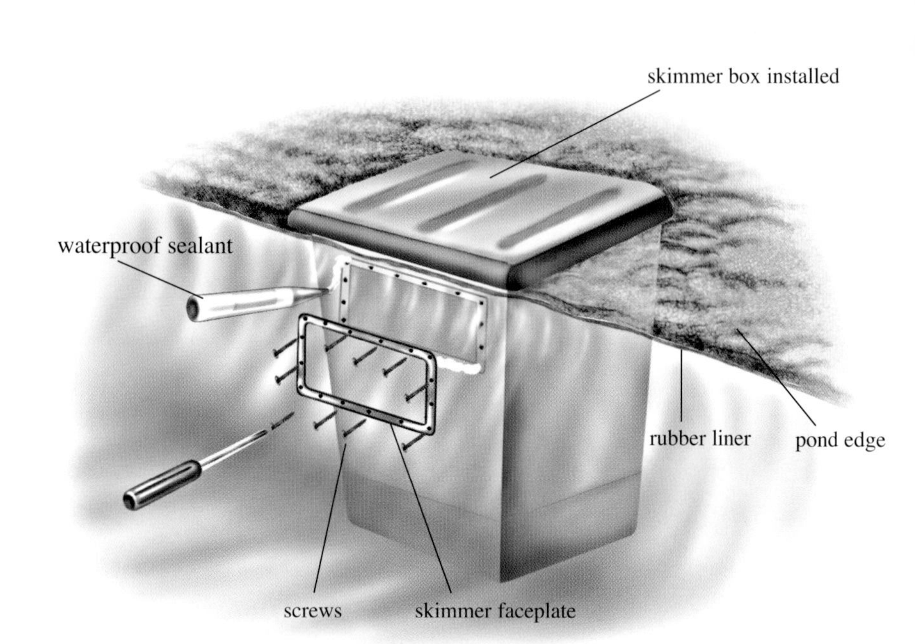

➊ Secure the liner to the skimmer box and prevent leaks by applying a watertight adhesive between the faceplate (secured here with screws) and the liner.

vent them from becoming clogged with dirt during the liner installation process.

Preformed liners can simply be dropped into the prepared space, though small adjustments may need to be made to allow the form to fit snugly. Liner options for larger ponds, however, may be significantly more difficult to install.

Underlayment and flexible liners often require pulling and tugging to get them to conform to the spaces for which they were intended. It is often helpful to gently step on the material, barefoot, to force it to fit snugly. Keep in mind that it may be necessary to repeat this entire process several times to ensure that the liner is draped evenly and allows for adequate overhang on all edges. The overhang should be sufficient to allow the liner to be fastened securely beneath flagstone, mulch, gravel, or whatever material will extend to the edge of your pond. As a general rule, allow at least 12in (30.5cm). It is better to overcompensate, allowing you to trim excess liner where it is needed.

Flexible liners may need to be trimmed and seamed in order to allow for total coverage of the pond bottom. Seam kits are available from a variety of pond product vendors and usually include a marine-quality silicone sealant designed to stand up to total submersion..

The same submersible sealant should be used to seal any openings in the liner that have been made to accommodate plumbing or mechanical pond components. Every cut and opening in the liner should be sealed thoroughly to prevent any leaks or weakening of the structural integrity of the pond. The first signs of leaks in garden ponds usually stem from old seams or openings in the liner that have not been sealed properly.

Custom concrete and fiberglass ponds present their own construction pitfalls. Their installation is not for the average do-it-yourself weekend warrior and should only be attempted with the help of a skilled contractor who has experience with similar projects.

BRICKS AND MORTAR

Pump houses can also be constructed to receive and conceal external pump and filtration equipment. Similar construction techniques are used to create partially raised and aboveground formal ponds.

Natural stone and cinder blocks are common choices for constructing exterior walls that make up elements of the formal garden pond. Natural stone is preferred for its aesthetic appeal, but cinder blocks are a less expensive alternative and are often used to construct the underground portions of pump houses or partially in-ground formal ponds. Cinder block can also be faced with less expensive cultured (faux) stone, brick, or other more attractive building materials.

Be aware that cement blocks and certain other building materials, when submersed, may leach chemicals into the water, which will make it inhospitable for pond life. Formal features can be lined with flexible liners for insulation against these harmful chemicals.

It is vital that the right building materials are used for your climate and outdoor conditions. Bricks and mortar are rated differently for indoor and outdoor use. Know which type you need before you head to the home improvement supply store.

Masonry can be a challenging undertaking. It requires special tools and equipment and can be daunting to the novice. If you're building a formal, walled garden pond that requires these construction techniques, be sure to educate yourself on the process from start to finish, and be prepared to ask a seasoned professional for help.

CHAPTER 8

Landscaping

Constructing, filling, stocking, and planting a garden pond is only half of what it takes to create your very own backyard retreat. One can't simply drop a garden pond into place and be finished with it. A masterpiece that will offer solace for years to come also requires careful thought and planning where landscaping is concerned. Creating the surroundings to bring out the natural beauty of your creation will elevate your garden pond to a work of art with maximum appeal.

When the contours of the surrounding earth complement the curves or lines of your water feature, the entire environment seems as though it belongs. Seating and viewing areas allow you to admire your handiwork and take in the relaxing benefits of your garden pond. Patios, decks, and gazebos provide gathering places for entertaining family and friends naturally drawn to the sounds of water. Terrestrial plants and nearby gardens intensify the sensory experience and reflect gracefully on the water's surface. Fencing or walls add privacy and charm to your new outdoor space. Paths and bridges encourage unhurried wandering.

Decide exactly where each of your landscape plants will be placed before digging holes.

SHAPING THE LAND

Berms, banks, and hills can shelter your garden pond from blowing debris and create visual interest by providing varied heights for planting. The shape, size, and elevation of these earthen structures largely depend upon individual taste. However, following a few basic principles can ensure you'll be pleased with the outcome.

If your garden pond's surroundings are too low and sprawling, it can make your feature look austere and unnatural. However, if you have chosen a more formal structure, this situation may be ideal. Similarly inappropriate, high banks and hills located too close to a pond can be a magnet for debris, channeling leaf litter into the water and crowding the garden, giving you a sense of claustrophobia. It may be necessary to step back frequently during the process of shaping the pond's surroundings, imagining the feature as it would appear with mature plantings.

Contouring the surroundings can be done by hand or with light-duty excavation equipment, depending on the degree of work needed. If you've chosen to install an in-ground pond, use dirt removed during the installation process. Otherwise, fill dirt and topsoil are readily available from landscape supply outlets and local quarries. Be sure to amend any areas you'll use for terrestrial planting with several inches of high-quality topsoil for optimal growth.

When building a stream such as this, it may be necessary to step back and examine your work frequently, trying to imagine the finished product and working to achieve the desired effect.

SIT BACK AND ENJOY

Providing space to gather and enjoy your garden pond is absolutely essential. Spend some time deciding the best possible vantage point from which to view your creation. Find that magical spot that offers the prime view of pond fish, moving water, showy plants, and all the things that bring relaxation and enjoyment. Then set out to create the most comfortable atmosphere for the type of benefit you want to get out of your water feature. If your goal is to come home after a long day and soak your feet in the cool water, place a flat flagstone or boulder near the water's edge. If you're planning on catching up with a loved one after a day at the office, consider a bench or patio furniture situated pondside, upon a level clearing.

➋ Make sure to create an area with seating, where you can relax and enjoy your completed garden pond.

Decks and Gazebos

Decks and gazebos offer excellent spots for lounging or entertaining, and constructing them is not as difficult as you might think. Many decks and gazebos are now readily available in all-inclusive kits at home improvement stores and lumber suppliers, making assembly and installation a snap. Many suppliers can even assemble custom decking kits, just for your situation. Simply prepare a list of the dimensions for your project, along with a rough sketch, and you'll walk away with everything you need, right down to the very last bolt.

BUILDING CONSTRUCTION TERMS

***Balustrade*:** A handrailing that caps a staircase or deck surround.

***Cantilevered*:** A construction technique that extends a structure, such as a beam, outward from a supporting fixed point—as in a deck or structure that hangs over water.

***Fascia*:** Face board used to cover supporting structures, such as a joist.

***Galvanized*:** Used to describe iron or steel coated with rust-resistant zinc.

***Girder*:** A beam used as a main horizontal support in a building or structure.

***Joist*:** Supporting structures that span walls or abutting girders (as with ceiling joist) and carry the weight of a roof or floor.

***Pier*:** Term used for aquatic construction, where a supporting post is anchored to the ground beneath the water line.

***Portland Cement/Ready-mix/Quikrete*:** Name brands of small-quantities of cement used to anchor supporting posts into the surrounding soil.

***Post anchor*:** Galvanized anchor attached to the bottom of a buried supporting post to keep it from shifting in the soil.

***Pressure treated*:** Wood treated to prevent rotting from exposure to moisture—also includes poison treating to prevent insect damage.

***Spindles*:** Structures connecting the balustrade to the railing footer boards.

Gazebos and decking can also be constructed to cantilever out over the water's surface to create the illusion of a pier and thrust you out into the garden pond itself. This method tricks the eye into believing that the pond extends forever, back beneath the decking.

When shifting the supports of these structures to allow for the water to lap underneath them, take care to securely anchor the posts. If the soil is soft, a simple post anchor may not suffice. You may need to cement and plumb these posts in place before securing subsequent joist, fascia, and deck boards.

Decks and gazebos make great additions to any backyard, especially when located pondside.

Patios

Patios are another popular option for pondside entertaining and gatherings. For patios made of pavers, flagstone, or cement slabs, the pond's edge can be allowed within just a few inches of the foundation, again, putting the viewer right in the middle of the action.

Installing patios using paving bricks or flagstone is not as daunting as it seems, provided there is a firmly packed substrate upon which to place the stones. If crumbly soil is present on the site, you'll need to stabilize it with crushed rock, brick, and building rubble. Four to six inches of this stabilizer material will prevent the patio from settling and heaving, which can force the stones out of alignment.

An added layer of sand, no less than 1-inch deep, creates a bed atop the solid foundation to receive the stones. Place several leveled strings across the patio bed for a visual guide that will help keep the courses of stone uniform and flat. Gently pound each stone into place with a rubber mallet, checking with a level as you go. Stone dust can be swept into the cracks between the stones to lock them into place and provide planting areas for creeping ground cover.

TERRESTRIAL PLANTS AND GARDENS

Aquatic and marginal pond plants make up only one facet of the possibilities at your disposal when creating a total environment for relaxation and enjoyment. Terrestrial plants and the gardens

FIVE STEPS TO A LOW-MAINTENANCE GARDEN

1. **Remove all weeds, sod, and sharp objects from the pla ing area.**
2. **Unroll enough of the barrier to cover the entire area.**
3. **Secure the edges of the barrier by forcing it down into the ground with the edge of a garden spade or shovel.**
4. **Decide where each of the plants will be placed, and cut-holes in the barrier large enough to accommodate the plants.**
5. **After securing each of the plants beneath the layer of weed barrier, cover the rest of the exposed barrier with pine or cypress mulch or decorative stone or gravel.**

Common landscape plants that grow near the water's edge and in nearby garden areas further enhance the appeal of a garden pond.

they create on the periphery of your garden pond can bring an added dimension of beauty to any water feature.

No matter what types of plants you decide to incorporate in your landscaping, it's important to keep in mind what your environment will look like when the plants are fully mature. Most plant care labels will list the plant's maximum height. When placing your plants, be sure to leave enough room between specimens to accommodate their size. This will help prevent the need for drastic splitting and pruning that can shock or kill your plants.

Given that a garden pond needs a certain amount of continual maintenance to ensure stable water quality, many people are reluctant to expand the garden to the surrounding environment. The fear is that weeding and maintenance of the grounds will simply be more than they can handle. However, with a few basic steps, you can anticipate and prevent many problems and enjoy the best of both worlds, without the hassle of increased time and effort.

The advent of fabric weed barrier, now widely available at most garden supply stores, has meant an end to gardens overrun with weeds that require constant attention. The fabric, usually black, allows water to penetrate down into the soil but prevents weeds from receiving the sunlight they need to grow and reach the surface. Rolls of weed barrier, also available in semipermeable plastic, are well worth the expense and application.

The barrier, which lasts for several seasons, cuts down significantly on the amount of maintenance necessary. Weeds often appear growing in the mulch or gravel on top of the barrier, but, since there is no top soil for them to root into, they lift off the planting bed with ease.

Trees and Their Roots

Roots are indeed an issue, as they reach above or lurk below the surface of your garden pond. People are often reluctant to locate garden ponds near trees or to plant new trees anywhere near a water feature. Most of these fears stem from witnessing a tree root that has buckled a sidewalk or road surface. The gut reaction for most people is to automatically assume that a tree will search and destroy any water feature within reasonable distance, puncturing liners and breaching solid walls with its powerful roots.

It's really a question of plant biology. Trees are hardwired to search out moisture for survival. When a small crack develops in asphalt or concrete, a tree root heads north in search of the water pouring in from above. However, if a garden pond is installed properly, the surrounding ground should be bone dry, meaning tree roots will look elsewhere for sustenance.

Another concern about trees involves the added plant waste they can drop into a garden pond. Spring blossoms and fall leaves or a continuous shower of evergreen needles are sure to find their way into your water feature. This shouldn't be a problem if you're prepared with adequate skimmer and biological filtration components. You may simply have to empty the skimmer basket more often than the manufacturer recommends.

Planting for sensory appeal is largely subject to personal taste and growing zone. Whether potted and placed on the capstone of a formal feature or in beds surrounding an informal pond, with a couple of basic considerations, terrestrial plants offer an unlimited pallet of creativity.

Sensual Environment

Planting to appeal to the senses can truly intensify the sense of relaxation induced by a garden pond. From color and texture to touch and scent, surrounding plantings can truly transport you to another place and time.

Color. Choose plants with a variety of colors. Terrestrial plant foliage comes in a variety of hues from light green to crimson red to variegated combinations. Planting with splashes of color helps bring life to the area surrounding your garden pond. For instance, a dark corner can be transformed into an environment that actually seems lighter by planting a bright, variegated specimen. Be sure to consider sunlight constraints posed by existing trees or mature shrubbery.

Blooming terrestrial plants are among the most popular around the garden pond. With proper planning, these plants can offer color year-round. Perennials, annuals, and shrubs show their stuff only under the right conditions. In spring, for instance, the bulbous plants such as crocus, daffodils, and tulips, as well as flowering trees, burst with fragrant blossoms in every color of the rainbow. Summer offers annuals such as foxglove, impatiens, marigolds, and a host of others a chance to shine. And fall unleashes the hardy chrysanthemum. Not to mention winter's evergreens with colorful berries, such as holly, winterberry, and Japanese yew. Foliage from the previous seasons can provide striking off-season beauty. Choose a variety of plants from each of the categories above, place them in varying locations in your garden pond environment, and you'll be in for a visual treat every month of the year.

➊ Appeal to the visual sense by including trees and other plants with vibrant colors, as in this garden pond.

Texture. Planting for texture also adds flair to an ordinary space. For instance, by pairing delicate, lacy ferns with strong, broad-leafed hostas, you can transform an average shady pond periphery into a sensory experience of its own—one that springs to life in the slightest breeze.

Planting for height around the garden pond can show off its best attributes and provide the perfect backdrop. Pampas and other popular tall grasses make a space seem bigger by drawing the eye upward. They create a screen from the wind and nearby

➔ Lavender adds texture and long-lasting color to a garden pond environment. It also adds therapeutic scents believed to heal and relax.

neighbors with their dense blades. This also acts as a solid field to view your garden pond against.

Aromatherapy. Planting for aromatherapy around the garden pond can further enhance your enjoyment. Planting for scent in the garden is a time-honored tradition first practiced more than 6,000 years ago by Persian royal gardeners, who, incidentally, also were among the first to build garden ponds. Fragrant plants often release their scents when touched, making their enjoyment a truly multisensory experience.

Garden aromas have been thought to offer a variety of curative benefits, according to believers in the efficacy of aromatherapy. Lavender, known for its relaxing and healing properties, offers the added benefit of delicate, long-lasting purple blooms. Although sage does not offer much of a visual punch, its strong cleansing and euphoric odors make up for what it lacks. Mint,

➔ Creeping thyme grows and spreads easily between the cracks of stone pavers and flagstone. Light foot traffic causes the plant to release its relaxation-inducing essential oils.

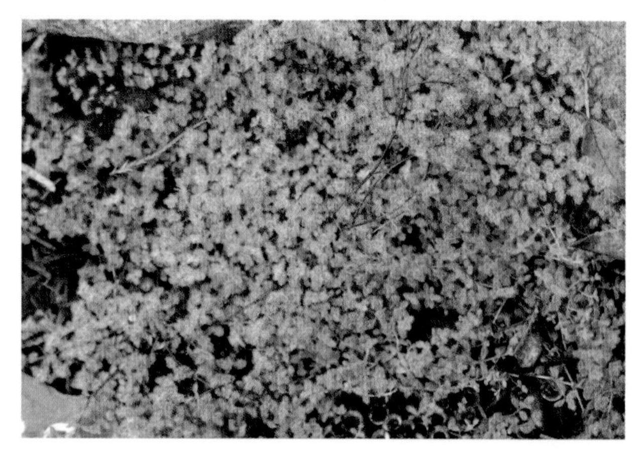

largely unnoticed in the garden with its tiny purplish flower cones, is known the world over for its soothing properties. The faint citrus smell of scented geraniums and lemon thyme lingers, producing an atmosphere of well-being popular with many hard-working commuters. Creeping thyme, which grows exceptionally well between the cracks of flagstone patios and pathways, releases its delicate aroma when stepped on. Roses and other aromatic flowering plants can also add to the scent experience in a garden pond setting. See Table 6 for a list of some common garden plants and the feelings and emotions they are thought to induce.

TABLE 6 PLANTING FOR AROMATHERAPY

Plant	Feeling/Emotion
Bergamot	Mood elevating
Chamomile	Relaxing
Frangipani	Relaxing
Lavender	Healing
Lemon thyme	Mood elevating
Lily	Euphoric
Mint	Relaxing
Rose	Euphoric
Sage	Cleansing
Scented geraniums	Mood elevating

FENCES AND WALLS

Garden fencing and walls create the sense of an outdoor living space with a defined border that can be perfect for adding privacy to a suburban setting or character to a country garden pond. Fencing is available in a wide range of materials from willow sticks to bamboo (in premade rolls) to more traditional privacy or picket styles (available in premade sections). Walls can be fashioned out of inexpensive cinder block and faced in cultured stone, brick, or exterior tile, allowing for maximum creativity. Or they can be made with more traditional materials, including concrete, field-

➔ If you live in an area where open water is considered a hazard, you will probably be required to surround your pond with fencing. Or you may choose to include fencing as part of the overall design.

stone, or wall blocks. Regardless of which material you use to separate you from your neighbors, you may have no choice but to include the construction of a fence or wall in your master plan.

Your garden pond may be legally considered a pool, depending on your municipal codes, which also may require you to include a fence or wall surrounding the garden pond. In the interest of safety, you may want to include some sort of enclosure anyway. Should an unattended toddler wander into your garden pond, you could have the recipe for disaster. Whether tall, slender bamboo to match a Japanese water garden theme or a fieldstone-and-mortar wall to accentuate a more formal feature, choose a material that best matches the overall impression you would like to achieve.

PATHWAYS AND BRIDGES

Pathways and bridges invite you to wander throughout the gardens, enjoying your pond from every angle. Pathways are essentially modified patios. Although they can be made of crushed stone or gravel, using the same stone as that of a nearby patio will

SAFETY ALERT

A toddler can drown in as little as an inch of water. In most municipalities, artificial bodies of water deeper than two feet are classified as swimming pools and have to be surrounded by fences. Most water features, therefore, are not required by law to be fenced in. However, given the potential for a safety hazard, you may want to find a way to protect yourself and your neighbors.

➔ Welcoming pathways, such as this one, draw visitors into the garden and toward the magic of a garden pond.

help unify the elements of your garden pond. The ground can be prepped in the same way for both applications and installation is essentially the same.

If you plan to enjoy your garden pond after twilight, an evening stroll by low-voltage pathway light can be the perfect end to a stressful day.

Similarly, garden pond bridges can be made from modified decking. The principles for anchoring the supports, securing joists, and laying deck boards are the same. However, you'll want to avoid spanning more than eight feet of open space (a good rule of thumb for all pond bridges; otherwise you'll need additional support in the center of the span). Avoid the use of treated lumber, as it may contain chemicals harmful to plants and pond fish.

➔ Bridges come in many forms, such as this single piece of sculpted material used to span an edge of a garden pond.

➔ Bridges may present special challenges, like the underwater piers that were needed to support the wide expanse of this structure.

Instead, use rot-resistant wood, such as redwood or cedar, or plastic composite material that will hold up to the exposure to added moisture.

There are several other options for spanning the shores of your garden pond. The most elementary and ancient method uses a single slab of stone. Stone monument makers and stone suppliers often have these long spans that can be simply placed between the two shores. The only drawback is the tremendous weight of these monoliths, which require heavy equipment to place. This

problem can be solved using smaller stepping stones, placed atop stacked bricks or rocks beneath the water's surface. You can also use stones end-to-end or zigzag longer stones from one side to the other. If working in a flexible liner pond, be sure to place extra padding beneath the stacked piers that support these structures to avoid punctures under the added weight.

↑ This bridge—nothing more than a large paving stone—seems a natural extension of the path on either side, imparting a sense of fluid movement to the overall design.

CHAPTER 9

Stocking Your Pond

Adding fish to the garden pond environment brings elegance and true beauty into the equation. The graceful swimming of koi, goldfish, and other species can bring an absolute level of inspiration and relaxation found nowhere else. Their bright colors delight and amuse the senses. Similarly, pond plants add a remarkable amount of color and texture to the garden pond. They add shade and food for pond fish and offer a feast for the senses for pond keepers.

STOCKING AND CARING FOR FISH

The decision to introduce fish should not be taken lightly. Adding pond fish completely changes the level of care your pond requires, as well as the delicate balance of the watery environment you have created. Armed with the right education and commitment to care for these creatures, it can be an exceptionally rewarding experience.

Which Fish Are Right for Your Pond?

Your geographic location, the time you can dedicate to maintenance, local laws and regulations, and personal preference are just a few factors that might influence your decision to add pond fish. Regardless of the species you choose, careful consideration of these factors will ensure that you have a positive and enjoyable experience.

Popular standards. Goldfish and koi are hardy, colorful standbys that are popular with most garden pond keepers. They can tolerate swings in temperature and water quality that would injure or kill most other ornamental fish. Koi are available in a wide range of color patterns, sizes, and prices. They are commonly thought of as the jewels of the realm of garden ponds. Goldfish, including varieties such as shubunkin, comet, calico, oranda, celestial, fantail, and moor, are also popular and sturdy fish for the beginner.

Other options. Several other varieties of colorful fish native to temperate regions of Europe are finding their way into garden ponds as well. Their behaviors, appearances, and colorations make them attractive options. Just be wary of your local legislation, which may prohibit the shipment or keeping of these fish in the backyard environment. Check with your local fish and wildlife office before choosing these or any other lesser known garden pond fish.

Orfe, available in silver and gold forms, is a muscular schooling fish that thrives when kept in a social grouping and feeds on surface insects, occasionally jumping clear out of the water to catch passing insects. Rudd, another surface-feeding fish available in gold and silver, also swims actively and tolerates a wide range of temperatures. Another tolerant schooling fish is the tench, which is available in gold and spends most of its time cleaning waste and detritus from the pond bottom.

Wild natives. Fish native to your area are often very accustomed to your particular climate and can be excellent additions to a garden pond. Many of them double as biological mosquito controllers. Gambusia, marketed as mosquito fish, and three-spined stickleback (native as far north as New Jersey) are sometimes available free of charge to garden pond keepers from local vector control authorities, who encourage their use in backyards. Other fish, such as rosy red and shiner minnows, are available as bait but liven up a garden pond as they travel in flashing schools. Catfish and bullheads also serve as cleaner fish and can be caught in many local waterways.

Game fish. Game fish, often found in local lakes and streams, can also be welcome additions to backyard ponds, depending on your preferences. Though these fish are not typically as brightly

Hardy goldfish are an aquarium standard but thrive just as easily in an outdoor setting.

colored as ornamentals and are secretive by nature, many garden pond keepers enjoy their antics. Game fish should not be kept with ornamentals, which they will consider bait. Hardy game fish, including bluegill, sunfish and black crappie, can be caught, kept, and returned when they outgrow your pond. Other fish, such as pike and bass, need larger, earth-bottomed ponds and are typically unsuitable for the average garden pond.

From the aquarium. Oscar, plecostamus, killifish, zebra danios, mollies, and several other vigorous freshwater aquarium species can be kept in garden ponds, provided the pH and temperature requirements for that species are maintained. Frequent monitoring of both variables is vital when experimenting with these fish. As water temperatures dip below acceptable ranges, these species will need to be brought indoors to protect them. Be prepared to shuttle them back and forth as warranted by seasonal temperature swings, unless of course your garden pond is located in a tropical climate.

Where to Buy Your Fish

Choosing the right fish to add to your garden pond depends on several factors. One of the most important is your own geographic location. If, for instance, you live in a suburban area near a major metropolitan center, your choices of retailers and specialists may be numerous. But if you live in a rural area, far from big-city life, you may have a hard time finding even one reputable source from which to purchase your fish.

The most commonly purchased pond fish are koi, known for their showy ornamental color patterns and hearty ability to survive outdoors in the garden pond environment. Although these fish are often available at pet stores and aquarium fish outlets, their care is specialized and best managed by a commercial importer or koi specialist.

Buying from a dealer. Dealers who stock only pond fish will ideally have a total understanding of the needs of the fish they offer. Generally, these dealers are found in major metropolitan areas and may be just a short drive away. A dealer will be able to tell you where the fish originally came from; how long they've been quarantined to isolate illness; and which size, number, and variety of fish are right for your environment.

Disreputable dealers do, however, exist. So the most important thing to consider is the overall health of the fish from which you are choosing. You want to be sure the dealer isn't selling deformed or sick fish and that the fish you are considering have not been housed with other sick fish. Similarly, when purchasing koi and other pond fish from an importer (as many koi are

Colorful koi, such as these, can be found in a wide variety of sizes and color patterns, making them a popular choice for pond fish.

imported from Japan), it is important to ensure that the fish appear generally healthy and have been quarantined properly.

Buying from a pet shop. Not to say that pond fish should not be purchased from pet stores. It is simply less likely that the staff will be able to answer specific care and health questions or will have taken proper care of the fish they are merchandising. The danger of purchasing pond fish at one of these general retail outlets is simply that the person making the sale could be a dog groomer, reptile keeper, or at the very least a general nonspecialist who looks at the fish as simply fish and isn't aware of their basic needs and requirements. However, a well-established business with a long history of happy customers is not to be overlooked. Visit these stores and develop your own opinions of the fish in their care.

If your options are limited, a pet store can be a good option for purchasing your pond fish. Many diamonds in the rough have found their way into holding tanks at pet stores. Look closely, and you just may find yours.

Buying online. It's best to beware of purchasing pond fish sight unseen. In these instances, you are simply not able to inspect the facilities yourself or determine the general health of the fish. Any number of things can happen during the shipping process. These usually hearty fish can become stressed and sick during long, turbulent voyages, just as humans are susceptible to jet lag and illness on long flights.

Nevertheless, when choices are slim, working with a seasoned professional dealer who has a history of shipping fish successfully can pay off. Risks can be lessened if the dealer offers a solid guarantee, which can cover a number of mishaps during and shortly after the shipping of his or her stock.

What to Look for

Regardless of where you decide to purchase your pond fish, there are several things to look for on a shopping trip. They all require that you look past the flashy colors and graceful appearances to the smallest of details that can speak volumes about the health of the fish you are about to purchase.

Start by asking the retailer a few very important questions, such as: Where did this fish come from? How long has it been in your tank? For how long was it quarantined? Has it ever been treated with any medications?

Continue by carefully observing the fish's physical appearance. If possible, ask the dealer to feed the fish in the stock tank and observe how the occupants react, allowing you to also observe the fish's swimming and behavioral patterns. Fish should appear healthy and actively interested in food.

Table 7 lists common abnormal observations and the illnesses they indicate. Avoid purchasing pond fish presenting these following warning signs.

➔ When shopping for fish, check for signs of illness, such as the cloudy eyes of this koi.

TABLE 7 SYMPTOMS OF POND FISH ILLNESS

Observation	Possible Illness
	Parasites
Loss of appetite	External parasitic infestation
Lethargy	External parasitic infestation
Bursts of swimming	External parasitic infestation
Scratching	External parasitic infestation
White or gray, fuzzy film	External parasitic infestation
Gray/white threads protruding from beneath scales	External parasitic infestation
Frayed fins	External parasitic infestation
Gasping for air at surface	External parasitic infestation
Listlessness	Internal parasitic infestation
Weight loss	Internal parasitic infestation
Slimy white feces	Internal parasitic infestation
	Bacterial Infections
Cloudy eyes	Fin rot
Red fins	Fin rot
Bloated abdomen	Dropsy
Loss of appetite	Dropsy
Congregating near waterfall or jet returns	Gill rot
	Viral Illnesses
Open reddened sores	"Hole in the side" (ulcer disease)
Smooth waxy lumps on pectoral fins or the lips	Carp pox
Hard, fine nodules or lumps	Lymphocystis

You Get What You Pay For

The price of pond fish, particularly koi, varies dramatically. Adult, show-quality koi specimens can cost tens of thousands of dollars or more. Rest assured; for the hobbyist, there is a wide continuum to suit any budget.

➔ Add life and color to the water with brightly colored koi such as these.

Price depends on several factors—the most obvious of which is size. The bigger the fish, the longer it has been cared for and the more money someone has invested in its health and happiness. That translates into a larger cost at the register.

The fish's location of origin contributes to its price as well. For example, a koi imported from Japan has come a long way to get to your garden pond. Shipping costs increase the retail price considerably, as do the higher quality of the imported fish. As a rule, fish from Japan are bred from pedigree lines of fish over several generations, specifically for more vibrant colors, and are more likely to pass on their genes to progeny produced by unrelated breeding. Again, more investment means higher price.

Color patterns also affect the price of koi. There are numerous, intricate color patterns present on the bodies of koi. The location, symmetry, and color of these markings determine the rarity or desirability of the fish. The more vibrant or intricate the coloration and pattern, the more expensive the koi.

This of course leaves room for flexibility in pricing, as much of what determines the value of koi is a matter of individual interpretation. For this reason, it's not uncommon to bargain for the price of a larger, more valuable fish.

Avoid the temptation to buy out the store. It's important to remember restraint, especially for beginner hobbyists, who typically lose several fish in the first year of pond keeping. Starting with small numbers of small fish can help avoid devastating losses to the garden pond and the pocketbook.

Fish under six inches are less likely to survive the first year in a pond. They aren't as well established and haven't been exposed to a wide range of water qualities or diseases. But if you're planning to start small, remember young koi can be 4-8in (10–20cm) during the first year, increasing in size by as much as 8-12in (20–30.5cm) during the second and several more inches during the following years. These fish can eventually reach 24in (61cm) or more in length, averaging an adult weight of eleven to twenty-two pounds. Consider how many fully grown fish your pond's filtration system can handle. You should also consider that your pond fish may breed without any help from you, resulting in complete pond overload, not to mention the fact that captive-bred koi can live for twenty-five years or more. Start small now to ensure success later.

Introducing Fish to Your Pond

Purchasing pond fish is only the first step in adding them to the environment you have created. Introducing new fish to a garden pond must be done carefully, with regard for the fish's safety and the safety of other fish that may already call your pond home.

Feeding time can be a great opportunity for interaction with pond fish, which, like these koi, can be trained to eat right from your hand.

Fish that are being introduced to a pond that is already stocked must be quarantined to rule out any medical problems and prevent your pond's inhabitants from becoming ill. Set up a separate hospital tank or container for the quarantine period, which should last for a minimum of four weeks.

Before purchasing fish be sure to ask the retailer the pH value and temperature of the water in the holding tank. You'll want to be sure it is as close as possible to the pH and temperature in your garden pond. Any more or less could send your fish into shock. A variance of as little as 10 degrees Fahrenheit could be fatal. Even before you bring the fish home, take steps to ensure the pH and temperature in your pond is very close to that of the store's tank. Add water buffers if necessary. Gently heat the water with a deicer or submersible birdbath heater; cool it by adding small amounts if ice made from pure water or cool, dechlorinated water. Try to equalize these factors as closely as possible before traveling home with your purchase.

Fish typically come home from the store in a large plastic bag. If the trip home will be a long journey, be sure to have the retailer fill the bag with oxygen before it is sealed. To avoid causing unwanted stress, the bag can be placed inside a cardboard box or paper bag to prevent direct sunlight from warming the water. When you arrive home, float the bag in your pond or quarantine tank for no less than 20 minutes to give the water on the inside of the bag time to equalize with the temperature on the outside.

Nutritional Requirements

Koi are omnivorous, meaning they consume both plants and animals. They need a mixture of both types of proteins in order to maintain good health. There is an endless variety of commercially produced pellet and flake foods that fulfill all of these needs. These diets can also be supplemented with treats of raw fruits and vegetables such as leafy greens, carrots, squash, watermelon, and citrus. Insects, earthworms, and other fresh or freeze-dried live foods can also be offered on occasion. Other pond fish, which are not known as scavengers or bottom feeders, may require higher levels of protein. Offering supplements of freeze-dried worms or feeder minnows corresponding to the relative size of your fish should accomplish this.

Fish are cold-blooded, meaning their metabolism fluctuates with the changing temperature of their environment. Therefore, the changing seasons and water temperature play a significant role in their care. Higher animal protein foods containing fish meal should be fed during the summer months to increase the fish's reserves in preparation for winter. High plant protein foods containing wheat germ are usually offered during the cooler months, as it accommodates the fish's slowing digestive system. The frequency of feedings and amounts offered vary throughout the year according to water temperature.

In extreme water temperature situations, a fish's metabolism goes into conservation mode. During winter temperature of less than 55 degrees Fahrenheit and summer temperatures in excess of 90 degrees Fahrenheit, metabolism of food stops and metabolism of stored reserves takes over. Under these circumstances all feedings should stop. Hardy pond fish, such as koi and larger goldfish, can last for several months on these reserves. However, if you live in a particularly cold or hot climate, and there is a danger of your pond exceeding these extremes for more than three months, consider removing your pond fish to a stabler indoor environment.

The amount of food that should be offered to your fish correlates directly with water temperature and total pounds of fish your garden pond contains. Note that fish require different frequencies of feeding throughout the year, as the water temperature fluctuates with the seasons.

A good rule of thumb is to feed your pond fish only as much food as they can consume in about 10 minutes. They will continue to eat whatever is offered but, because of their short digestive tract, will only derive nutritional benefit from a small portion of the food they consume. The rest will end up being expelled as waste. After 10 minutes, remove the uneaten food with a skimmer.

For a more precise measurement of how much food should be offered, try using Table 8 as a guide.

TABLE 8 POND FISH FEEDING CHART

Water Temperature	Percent of Total Weight of Pond Fish to Offer in Food	Frequency
Under 55°F	0	Stop feeding
55°F–60°F	0.5–1	Per day in 1 portion
60°F–70°F	1.5–2	Per day in 2 portions
70°F–85°F	2.5–3	Per day in 3 portions
85°F–90°F	1	Per day in 2 portions
Above 90°F	0	Stop feeding

FEEDING GUIDELINES

The guidelines represented in Table 8 break down into the following mathematical equation, developed by commercial breeders.

Formula for feeding pond fish in 70°F–85°F water:
Total fish weight in lbs x 0.03 = ___lbs of food
Total daily portion of food in pounds /3 times a day
= amount per feeding

Example: 10 lbs x 0.03 = 0.3 lbs. divided into 3 equal portions (0.1 lbs converted to ounces equals 1.6oz) for 3, 1.6-ounce feedings per day.

For example, a 12- to 14-inch fish weighs approximately 1 pound. If the pond contains 10 fish of that size, that would give you 10 pounds of total fish body weight for your garden pond. The recommendation is to feed your fish 3 percent of their combined body weight each day. Therefore, 10 pounds of fish multiplied by 0.03 equals 0.3 pounds, or approximately 1/3 of a pound. Because fish need to be fed three times a day during the warm summer months, that 1/3-pound portion would need to be divided into three equal parts and fed accordingly.

➜ A host of woodland creatures will be drawn to your mini-ecosystem—some harmless, others devastating. Learn how to discourage the fish-eating critters found in your area.

Predators

Be wary of predators that discover a smorgasbord in your garden pond. They can be notoriously difficult to deter. From cats to raccoons to fish-eating birds, they'll do just about anything to gain access to your prized pond fish.

Annoying sticky plants or thorny ones can discourage them from getting too close. There are also several products on the market

designed for this purpose, including electronic alarms and motion-detection sprinklers to frighten the wits out of critters and discourage their presence.

But winged assassins don't seem to mind as much and can be notoriously difficult to curtail. Decoys, such as owls, and other predatory animals, such as alligators or fellow bird species, such as the Great Blue Heron, sometimes work, but they must be moved often for the sake of realism.

To preserve the aesthetics of your garden pond, try lower-impact solutions such as stringing nearly invisible strands of monofilament fish line or fine bird netting across the pond's surface. Both will discourage hungry creatures from landing. A fail-safe strategy is simply to provide ample space for your pond fish to hide. Rock ledges and overhangs, deep pools, or submerged tubes and containers provide excellent refuge for a school of frightened fish.

↑ Netting may be the only thing that consistently keeps pond fish safe. The one drawback of netting is its appearance, which detracts from the aesthetics of the pond.

← Mechanical barriers, like chain link fencing, keep four-legged bandits from raiding this pond. But stands of wire or monofilament line may be needed to keep birds from landing and helping themselves.

POND PLANTS

Pond plants offer unlimited possibilities when it comes to adding life and beauty to the garden pond. The sky really is the limit where plants are concerned. From out-of-sight plants beneath the water line to border plants used to disguise the pond's liner edge, the number of choices you have is truly staggering.

Whether you want to create a natural, wetland environment with cattails, grasses, water lilies, and lotus or a more formal retreat with well-trimmed pondside shrubbery, careful consideration of the plants you'll use can tie the entire design together and create a garden pond you can be proud of. Keep in mind, if these plants bloom, that the blossoms will not last for more than a couple of weeks. However, proper selection can ensure some sort of color, and variegated or colorful foliage types can create beauty throughout the year.

➊ Smaller, more delicate blooms, such as the ones produced by this pink hardy lily, add a touch of grace to a relaxing oasis.

Getting the plants into the pond and keeping them healthy isn't as difficult as you might think. Submersed, underwater plants need only to be weighted down with minimal planting material and a few larger stones. Water lilies can be placed in pots made and sold specifically for them that allow for water circulation around their roots. Lotuses can be placed in any pot that contains no drainage holes. Marginals can be placed in slotted pots or net/fabric bags on shallow shelves with minimal fuss. And terres-

➋ Free-floating pond plants, such as this water lettuce, drift across the surface as their roots hang below, filtering nutrients and toxins from the water.

Special slotted pots designed to hold aquatic plants keep the roots firmly in place while allowing water to circulate freely.

trial plants simply go into the surrounding landscape to further naturalize the surroundings and create a true backyard retreat. The fact is, these plants are hardwired to do well in your pond with minimal effort from you.

Planting and Care

Plants are relatively low maintenance. They require three basic ingredients for survival: 1) light, 2) water, and 3) nutrients. Number one can be addressed when choosing plants and locations for planting. Place plants where they will receive the proper amount of light they need for survival. Number two is by far the

Potting underwater plants, such as these water lilies, instead of placing them in plant pockets in the pond itself, allows you to move them easily for repotting or overwintering.

easiest. Aquatic bog and marginal plants are self-watering. You need only worry about the pond's water level. Number three is a bit more difficult.

In gravel-bottom ponds, underwater plants can simply be tucked beneath the substrate or planted in preexcavated pockets. Most water gardeners recommend planting without traditional soil, which can add biological waste and foul the pond water. Baked clay, washed sand, and gravel mixes are available and marketed as aquatic soils and should be used when securing the roots of plants. Heavier gravel may need to be used to hold the roots in place and prevent the plants from floating to the surface. This also prevents pond fish from foraging for food in loose planting media, which can unpot the plants.

REPOTTING AQUATICS

Unpot the plant to check for swirled roots, rhizomes, runners, or tubers that indicate the plant is root bound and needs to be moved to a larger pot.

Select a pot no larger than two inches in diameter than the pot from which you are moving the plant—any larger may send the plant into shock.

Partially fill the new pot with just enough planting media to bring the crown of the plant to the top edge of the new pot.

Place the plant in the new, larger pot.

Add a time-release or aquatic tablet fertilizer to the pot.

Gently push new planting media into the gaps around the sides of the root ball.

Return plant to the pond.

Nutrients come in the form of aquatic fertilizers, available in time-release tablet form. These tablets should be added to each planting bed on a monthly basis throughout the growing season, as directed by the product's packaging.

In more formal ponds or in smooth-bottom ponds with planting shelves, pots will need to be used to hold the plants in place. There are two basic types of pots available for aquatic plants. The first, a slotted plastic pot, acts as a sort of cage, holding the planting media in place and containing the fertilizer tablets. The second, made of fabric, accomplishes the same thing. Both versions allow for root expansion and water circulation. However, hardy plants will need to be repotted or split each year to accommodate their increasing size.

Adjusting the plant to its proper planting depth should be done using stacked bricks or flagstone. Avoid using cement block, as it can change the chemistry of the water in your garden pond.

Trim and remove dead plant material and spent blossoms to keep your pond beautiful and chemically stable. Your plants'

↑ To say that water lilies are available in a variety of colors is an understatement. New varieties in stunning hues are hybridized and introduced to the consumer market each year.

← Many water lilies, such as this tropical red variety, feature blooms that rise up out of the water for an even more dramatic display of color.

Other floating plants, such as this unusual *Nymphoides*, which is marketed as water snowflake, can be found at your local nursery. Check with them often and you're likely to find an unexpected treasure to add to your garden pond.

needs may change with the seasons, particularly if you live in a cooler climate and plan on using tropicals or other plants you want to reuse after the frigid winter months. (See chapter 10: Seasonal Pond Care.)

Types and Quantity of Plants

How much biological plant material can your pond support and still maintain its natural balance? This depends on several factors, but assuming you have adequate biological filtration and your pond contains fish, there are guidelines that can help you decide when enough is enough. A good rule of thumb is to start the planting season with 30 percent of the water's surface area covered with plant material. By the end of the growing season, that surface area will have approximately doubled, provided the growing conditions are ideal and the plants have been adequately fertilized. Sixty percent of the pond's surface area pushes the limits of what a garden pond can support. At this ratio, plants can be trimmed or removed to ensure a stabler bioload.

If the choices and variables seem staggering, try breaking the plants down into more manageable groups. Each area and water depth of the garden pond environment is suitable for the growth of a specific type of pond plant. The majority of available plants fall into four basic categories.

Underwater plants. Underwater plants live out of sight and don't contribute to aesthetics, but they do offer a variety of other

SPLITTING OVERGROWN AQUATIC PLANTS

Unpot the plant and gently remove excess planting media.

Look for clearly formed divisions within the root ball, such as a daughter plant, tuber, rhizome, or runner with plants sprouting from it.

Using a clean, sharp garden spade or utility knife, sever any natural divisions. For more delicate tuberous plants, such as lotuses, you can simply snap them apart with your fingers.

Plant the cut end several inches down into fresh planting media that has been fortified with fertilizer.

Place the cutting in a shallower, warmer area of the pond to give it a good head start.

When the newly planted cutting begins to send up leaves, you can move it to its proper depth.

benefits. They filter toxins from the water such as fish waste and salts that contribute to algae growth. These plants also provide supplementary food for pond fish who nibble on their tender roots and foliage, thus keeping the plants pruned. And they add oxygen to the water during daylight hours. However, because these plants feed off that same oxygen at night, they are not recommended as a substitute for mechanical aeration if your pond contains fish that require dissolved oxygen twenty-four hours a day. These plants can be planted on the very bottom of the pond or at any depth, as long as it is beneath the surface of the water.

Lotus plants grow well in most climates and add the benefit of large, colorful blooms that drop away to reveal an attractive seed pod.

⬆ Dozens of varieties of bulbous plants, such as this purple iris, grow well when partially submerged in shallow pond water.

⬇ The duckweed shown here is a vital ingredient for creating a balanced ecosystem, but left unchecked, it can multiply rapidly and overtake the entire pond's surface.

They can be visible beneath clear water in plumes and swirls. There is no cut-and-dried answer for the amount of these plants one can add to a garden pond. They do not count in the 30 percent factor discussed in the section on controlling algae, and some hobbyists have experimented with covering the entire pond bottom with them—essentially creating a sea of green foliage beneath the surface.

Floating plants. Floating plants are the real showstoppers in most garden ponds. Water lilies and lotuses are often the main reason people decide to build a garden pond in the first place. They are, by far, the most popular floating plants used by garden pond keepers. Water lilies (*Nymphaea* spp.) come in two varieties—hardy and tropical. Hardy water lilies can survive during the winter months in northern latitudes, as long as the water does not freeze down to the level of the plant's roots. Hardies, as they are often called, are available in a variety of colors from white to peach, orange, and red.

Tropicals can be kept in the garden pond environment year-round in tropical zones, but most are planted as annuals in cooler zones and replanted each year. However, tropicals can overwinter indoors in an aquarium setup or in cold storage. This allows them to rest for several months before beginning the growing cycle again.

Tropical water lilies contain a blue gene that hardies do not possess, making them available in a range of lavenders, blues, and purples. Tropical lily pads often appear streaked or variegated in maroon with jagged edges. Tropical water lilies often appeal to busy water gardeners because several varieties bloom in the evening hours or just after dark and offer more opportunities for enjoyment.

Both types of lilies, as with lotuses, cover different amounts of surface area,

Plants, such as this diminutive variety of Siberian iris, often grow vigorously and spread when placed in the high-moisture environment of a garden pond.

depending on the cultivar. Some require only a few feet to spread their leaves and blossom, but others require 10ft (3m) or more to flourish. Water lilies grow from rhizomes and can be split apart and planted in various locations in the garden pond.

Lotuses (*Nelumbo* spp.) grow particularly well in northern climates, contrary to popular belief. Most survive through the winter in the same manner as hardy water lilies. They are relatively shade-tolerant and spread vigorously by sending out tubers that form a maze of runners. Lotuses can also be split and grown elsewhere in the pond, but they must be handled gently, as these roots are delicate and easily broken.

The blooms and leaves of lotuses typically rise up out of the water from a height of 2 to about 8ft (.06 to 2.4m). Opening in the early morning hours, the blooms of these popular plants usually close again before the morning rush hour comes to an end.

Marginal plants, such as these reeds, grow well when partially submerged and include an extremely diverse group of plants.

Lily and lotus blossoms usually last for about three days. Many cultivars open each day as a different color; for example, opening peach on day 1, orange on day 2, and deep red on day 3. Both species will continue to bloom through the growing season.

Free-floating plants. Free-floating plants, whose roots hang suspended in the water, also add elements of beauty and function to a garden pond. They differ from floating plants in that they are not potted or anchored to the pond in any way. They float freely from place to place. Water hyacinth, fairy moss, water lettuce, and common duckweed are some of the most commonly used plants. These plants filter toxins from the water and add oxygen to a pond's chemistry, much as submerged plants do. These plants require still areas of water to grow and multiply and can often take over the entire pond surface. They can be easily skimmed off the surface and kept thinned out.

Thinned plants need to be disposed of properly to prevent damage to the local environment. Although most of these free-floating plants are not hardy in cooler climates, they have multiplied unchecked in the wild in warmer, southern locations. Seal them tightly in a plastic bag with no water inside, and place them in a trash bin where they will be transported to a local landfill.

Marginal plants. Marginal plants fall into the general category of plants that can be planted in the shallows of a garden pond, partially submerged in water. Grasses, reeds, rushes, and a variety of other bulbous plants, such as irises and calla lilies, fall into this category. Mixing vertical, spiky marginals such as sweet flag and iris with low-growing, softer plants such as moneywort and varieties of arum creates a natural border that enhances the garden pond environment.

Combinations are endless and offer room for experimentation. Have a garden favorite? Try it as a marginal. Some of these plants grow as well in ponds as they do in dry flower beds. This adaptable group ranges in height from a few inches to more than eight feet. Many, like the common canna, are so varied and vibrant in their coloration that they compete for attention with even the most breathtaking water lilies and lotuses.

Bog plants. Bog plants like wet feet and are often planted around the damp periphery of garden ponds. They require either very shallow standing water or very wet conditions. Bog gardens are often created with leftover flexible liner or shallow earthen trenches made to collect rain water or overflowing pond water.

A wide variety of plants grow well in bog conditions, including commonly available bulbous perennials, milkweed, lobelia, primrose, mallow, and spirea, as well as most varieties of ferns. Many marginal plants also do well in these areas.

Bog plants are typically winter-hardy in most of North America and can be found growing natively in boggy and marshy conditions. These plants are often added to a garden pond to attract wildlife. Butterflies, moths, frogs, and turtles are often drawn to marshy areas when inviting plants are present.

Space Invaders

Beware of invasive species, plants that can multiply rapidly and overtake your garden pond, choking the life out of other plants you have chosen. Although great strides have been made to keep these plants out of the horticultural trade, many of them are still available and offered as water garden plants by unwitting nurseries. Before making your final selections, check with your state's branch of the United States Department of Agriculture (USDA) office for a list of "noxious weeds"—plants that are illegal to possess and could cause a threat to your local environment if they spread from your backyard.

In addition, consulting a local water gardening or horticultural club could save you the headache of choosing plants that can overrun your garden pond. These groups may know even more than your state's branch of the USDA about which plants could spell disaster. Water hyacinth (a free-floating plant with white blooms) in Florida and submersed milfoil species throughout much of the rest of the country have cost local governments vast sums of money to eradicate from waterways. Surprisingly enough, these plants are still readily available in the aquarium and water gardening trades and thus could find their way into your own garden pond if you're not careful.

It also pays to consult a cold-hardiness zone map to find out which growing zone your pond falls into. Growing zone maps indicate the average annual minimum temperature for any given location. Zone numbers are often represented on plant tags at the nursery. Knowing your zone number helps you decide which plants will be a good match for your garden pond. Growing zone maps are readily available at most greenhouses, garden clubs, and governmental agency Web sites, such as the USDA.

CHAPTER 10

Seasonal Pond Care

Garden ponds can offer relaxation and enjoyment throughout the year. Even in the northern winter, a landscape blanketed in snow and ice can be a thing of beauty. The peaceful environment of calm created by dormant plants and the slumbering fish beneath the icy surface still has the power to soothe and comfort the weary.

With a few maintenance to-do lists for each season, you can make sure that a vibrant ecosystem will thrive. Moving through the seasons comes naturally in the wild. In the backyard, the slice of nature you have created will need a little help from you.

SUMMER

The summer season means longer, warmer days for enjoying your garden pond. It also means higher water temperatures and vigorous activity of aquatic plants and fish, which will require a bit of effort to control.

Keep an eye on the plants in your garden pond environment. Thin out submerged plants and floating lily leaves as needed to avoid their rapid spread, as this can crowd out the rest of the life in your pond. But it is also important to allow a certain ratio of plants to grow on the surface to shade the pond below, or it can become overrun with algae that thrive on full-sun conditions. As discussed in the section on plants in chapter 9, start at the beginning of the year with 30 percent of the pond's surface shaded by surface plants. Start pruning when the ratio reaches 60 percent.

ALGAE HAPPENS

Algal blooms often plague water gardeners during the summer months. Although there are many chemical additives that claim to help combat the frequency of this problem or keep it in check altogether, the best weapons are easy to come by.

Shade a portion of the pond.

Remove the algae as it is discovered with a net or stick.

Integrate aquatic plants that filter nutrients from the water (water lettuce, water hyacinth, watercress, etc.)—thereby starving out the offending algae.

A certain amount of algae in a pond is to be expected. It's not necessarily a bad thing. Pond fish often benefit from grazing on it. But, left to multiply, it can drain the water's oxygen content and cause murky clouding.

The increased growth of plants in and out of the garden pond attracts a host of backyard pests, most commonly aphids, beetles, slugs, and snails. They are attracted to the moist environment you have created. By simply pinching the offenders or hosing them into the water for your fish to dine on, you'll preserve the health of pond occupants. More severe infestations may require nontoxic sprays or powders, such as Bacillus thuringiensis (Bt). Bt is a bacteria that infects and kills soft-bodied insect larvae.

Chemical runoff from a nearby lawn can also spell disaster during the summer months, as can grass clippings that foul the water. Take care when tending the surrounding area to avoid these easy mishaps.

Added summer sunshine also means increased rates of evaporation. Be sure to check your pond's water level frequently. Topping off the pond regularly will help plants stay submerged under the proper amount of water and prevent exposed liner from drying and cracking in the heat. Add an appropriate amount of dechlorinator to the pond, if you're using tap water treated by a local municipality.

As pond fish become more active and metabolize more nutrients, they will add increased waste and toxins to the water. Be sure to test pH, ammonia, and nitrite levels more frequently during this time of year. Adjust these levels using the methods discussed in chapter 3.

Decaying plant matter can also cause spikes in toxins and should be removed frequently. Deadheading

Summer is an active time of year for growth of pond plants and vigorous behavior of pond fish.

marginal flowers and clipping dead leaves throughout the season will foster a more balanced environment.

FALL

As summer cools into fall, the garden pond slips into dormancy. Although this is less of an issue in year-round warm climates, in the temperate zone and northern latitudes, it's time to prepare by tucking in your backyard for a nice long rest.

If there is a deciduous tree near your pond, you'll need to be extra vigilant during this time of year to make sure excess leaves don't find their way to the bottom. During the winter months, these leaves decompose and give off lethal toxins that can be deadly for pond fish.

Be sure to check filter and skimmer baskets daily during this time of year. They will require frequent cleanouts to keep your equipment functioning properly, which should keep the leaf litter to a minimum. If your pond doesn't use filtration or skimmer components, you'll need to skim leaves off the surface with a net and retrieve waste from the bottom with a soft plastic rake or pond vacuum.

Fall means trees slip into dormancy, too, dropping their leaves into your garden pond.

You may want to consider tenting or covering your garden pond in netting to prevent blowing material from entering the water. Other methods include stringing a line of temporary fencing or staked netting along the windward side of the pond.

As the plants in and around your pond slip into dormancy, you'll need to cut away the dead plant material. Marginal plants can be trimmed to within an inch of the water's surface. Hardy aquatic plants can be cut to within an inch of the pot or soil's surface. Tropical plants can be discarded when they begin to die off or they may overwinter in a heated greenhouse or indoor environment. You may want to leave tall grasses and cattails through the winter months for visual interest.

Aquatic plants should be moved to the lowest point of the pond. If the water they are in is expected to freeze solid during the winter months, they should be removed to overwinter in cold storage.

If you are planning on shutting down your pond's pump or other mechanical components during the winter (a good idea if you live in an area with subzero temperatures), late fall is a good time to remove, clean, and store them. Even if your pond contains fish, you may need to shut down these components to protect them from ice damage. There are other methods for ensuring your pond water stays healthy through the winter, as we will discuss later. Be sure to open any check valves in your plumbing system to drain the pipes to prevent cracking and breakage from ice formation.

COLD STORAGE BASICS

Remove the soil or growing media from the roots of your lotus and water lily tubers. Other hardy perennials can overwinter in the ground.

Trim off remaining foliage and wrap them in damp newspapers or submerge them in containers of washed moist sand.

Place them in a cool dark place. A root cellar or refrigerator drawer works well.

Check them frequently and keep them moistened throughout the winter season.

Carve out any signs of mold and change newspapers or sand at the first sign of decay.

WINTER

The decision about whether to completely shut down your garden pond during the winter months depends largely on the climate in

As winter and its frigid temps move in, a pond's needs and those of its fish change.

➔ Hardy pond fish can survive through the winter, beneath ice and snow, provided there is still liquid water in the pond and open water, which allows harmful gasses to escape.

which you live and the pump your pond uses for water recirculation. Without a space of open water, pond fish can easily perish from lack of oxygen and toxic gas buildup from decaying plant matter. Ponds that remain under total cover of ice and snow for long periods of time require a certain amount of winterization. In warmer climates, winter is of little concern, but as the chill of winter sets in elsewhere, there are a few strategies for keeping your garden pond safe beneath the ice and snow. There are several methods that will make sure that your garden pond's inhabitants spend the winter months in peaceful slumber.

You may want to move your pond fish indoors during the cold winter weather. Be sure you have an aquarium that is large enough to accommodate your fish; you may want to consider building a temporary indoor pond from cinder blocks and flexible liner. Just be sure to properly acclimate your fish. Equalize the indoor environment to match the outdoor pond as closely as possible—especially with respect to water temperature, pH, hardness—and, of course, eliminate harmful nitrates and ammonium compounds. Introduce the fish in floated tubs, gradually adding water from the indoor pond, watching for signs of shock or stress.

If you live in an area with subzero temperatures that remain below freezing for weeks at a time, it's a good idea to take your garden pond offline for the season. Similarly, if you have chosen to construct a raised pond, you are likely to experience a solid winter freeze unless you are in a frost-free zone.

In order to shut down an inground pond with fish, you'll need to make sure there will be an opening in the ice to allow for gas exchange and for some form of moving water to provide aeration. As long as the pond doesn't freeze solid, the fish should be able to overwinter safely. A commercially available electric pond deicer will keep a hole open on the pond's surface. Submersed aerators or pumps with upward facing outlet jets break the surface of the water and provide the agitation needed to introduce oxygen. Shutting down a raised pond in a subzero winter zone involves draining it to ground level.

Some ponds, however, continue to operate during the winter months, despite their locations in frigid climates. Some people actually prefer maintaining their ponds through the winter months, enjoying the beauty flowing water and ice create. This is possible in ponds that use larger pumps that are able to keep the waterfall and stream elements of a garden pond moving rapidly enough to prevent the formation of solid ice. Typically, pumps that provide a flow rate of at least 2,000 gph are capable of supporting a pond through the winter months.

However, care must be taken to avoid ice dams. These can form in slow-moving streams and divert water up and over the edge of the pond liner. As long as the water continues to flow through the pipes, ice will not form in them. The only real danger is that if a power outage occurs and lasts long enough for the water in the pipes to slow and freeze, the entire system could be ruined or frozen solid well into spring.

SPRING

The time for awakening the garden pond brings with it chores that must be performed to ensure a vibrant, enjoyable season ahead. When the ice breaks and the water begins to warm, it's time to get to work.

Every spring should begin with a careful inspection and a little housekeeping. Check the pond structure for signs of settling, buckling, cracks, or tears in the liner or walls of your pond. Make

Spring cleaning often involves partial drainage and the use of a pond vac such as this one, which removes sludge and debris that has accumulated over the course of the year.

➔ Total drainage may only be required every three years or more. It may never be warranted if the pond is constructed well and not overstocked with pond fish and plants.

any necessary repairs. Test your GFCI to be sure it is operating properly. Net and inspect pond fish for illness (using the signs of illness chart found in the chapter on stocking your garden pond). Clean and reinstall equipment, and test it to be sure it is in good working order.

Note the overall condition of the pond after its winter dormancy. If it appears much as it did the previous spring, you can proceed with business as usual. If it looks fouled with sediment and smelly sludge, it may be time for a spring cleanout. A total cleanout may not be necessary every year. Depending on the size and bioload of the pond, it may only be called for once every three to five years.

A total cleanout involves draining, washing, and refilling your pond. Start by using a siphon or submersible sump pump (enclosed in wire mesh to avoid sucking up pond fish) to drain the pond. Channel the water to a nearby flower bed or vegetable garden. Pond water is an excellent fertilizer. If your pond features a bottom drain, simply open it and allow the water level to drop.

Wait to remove the fish until the water level is 6in (15.2cm) from the bottom. It will be easier to catch them

SPRING CLEANING CHECKLIST

- **Remove livestock**
- **Drain**
- **Clean pond and electrical parts**
- **Refill**
- **Dechlorinate**
- **Test water**
- **Acclimate and reintroduce livestock**
- **Recharge beneficial bacteria**

and less stressful for both of you. Carefully net the fish and place them in containers filled with water from the pond—many hobbyists use inexpensive plastic wadding pools with supplemental aerators to ensure the fish have sufficient oxygen for survival during the cleanout phase. The water taken from the pond to fill the holding containers should have the same water chemistry and temperature, making the transition safer and less stressful for the fish.

Aquatic plants can be removed and placed in the shade under dampened newspaper for a few hours while the cleaning is underway.

Place the pump or siphon at the lowest point of the pond and continue draining the water while gently spraying down the sides of your garden pond with a garden hose and spray nozzle to remove any sludge or floating algae that have clung to the sides during the draining of the pond. Try to avoid scrubbing or hard streams of water, as this will dislodge beneficial bacteria clinging to the sides of your pond. When the water from the hose, which collects at the bottom of the pond during the cleaning process, begins to run clear and the sediment has been removed, you can begin to refill the pond.

If you're using tap water from a city supply, be sure to add an appropriate dechlorinator. As the pond fills, place the containers holding the pond fish back into the water. (Doing this will slowly equalize the temperatures between the two environments and minimize the danger of shock.) Check the pH of the new water, as well as that of the water in the containers holding the fish. Take care to equalize both environments. When the temperature difference between the pond and the containers is less than 3 degrees, you can begin dipping water from the pond into the containers to gradually complete the acclimation process. After several minutes, it is safe to release the fish into the cleaned pond.

Aquatic plants can be carefully replaced and adjusted to the proper height with stacks of bricks. The beneficial bacteria will also need to be recharged after the cleaning to ensure a healthy, balanced ecosystem. Commercially available culture starters will help speed the process.

If a total spring cleanout is not needed, simply clean out accumulated debris from the pond and surrounding area, preparing the soil and water for a burst of new growth from below. Water lilies and lotuses can be moved up from the depths of the pond to the shallow areas. Place the pots 6–8in (15.2–20.3cm) from the surface to allow new growth to flourish in the sunny, warmer water. As the plants start to leaf out, they can be lowered to their recommended depth.

By mid-spring, check the roots of all potted plants. If they have outgrown their pots, divide the root balls and replant them. Share the extras with friends or with water gardening club members. Marginal and terrestrial plants can also be split and redistributed at this time.

Check your hardy aquatic plant tubers in cold storage and introduce them when the water temperature hits 50°F (10°C). Wait to introduce tropicals until the water reaches 70°F (21°C) and the danger of frost is far behind.

Throughout the year, with close attention to water quality, good maintenance practices and appropriate filtration and pump components, a garden pond is sure to offer hours of enjoyment and beauty. Whether sipping lemonade in a hammock at the water's edge on a Sunday afternoon or unwinding with your feet up after a hard day at work, with colorful pond fish swimming up to greet you, you're sure to find the solace and relaxation that are constant benefits to installing one of these treasures. Don't be stymied by the seeming technical nature of maintaining a garden pond. With minimal investment, you're sure to gain enormous returns and an enjoyable hobby for life. Born of good information, planning, creativity, and a desire for the good life, your garden pond will surely succeed and prove to be just the tonic for a work-weary soul. Take the leap; you'll be glad you did.

Appendix

Useful Conversions

U.S. Units to Metric

Inches	Millimeters
1/16	1.59
1/8	3.18
1/4	6.35
3/8	9.53
1/2	12.70
5/8	15.88
3/4	19.05
7/8	22.22
1	25.40

Inches to Centimeters

Inches	Centimeters
1	2.54
2	5.08
3	7.62
4	10.16
5	12.70
6	15.24
7	17.78
8	20.32
9	22.86
10	25.40
11	27.94
12	30.48

Feet to Meters

Feet	Meters
1	0.30
5	1.52
10	3.05
15	4.57
25	7.62
50	15.24
75	22.86
100	30.48

Volume Conversions

U.S. Unit	Metric
cubic inches	16.387 cubic centimeters
cubic feet	0.0283 cubic meters
cubic feet	28.316 liters
cubic yards	0.7646 cubic meters
cubic yards	764.55 liters
square inches	6.4516 square centimeters
square feet	0.0929 square meters
square yards	0.8316 square meters

Metric Units to U.S.

Millimeters to Inches

Millimeters	Inches
1	0.04
5	0.20
10	0.39
15	0.59
20	0.79
25	0.99

Centimeters to Inches

Centimeters	Inches
1	0.39
5	1.97
10	3.94
25	9.84
50	19.69
75	29.53
100	39.37

Meters to Feet

Meters	Feet
1	3.28
5	16.40
10	32.81
25	82.02
50	164.04
75	246.06
100	328.08

Volume Conversions

Metric Unit	U.S. Unit
cubic centimeters	0.061 cubic inches
cubic meters	35.315 cubic feet
liters	0.0353 cubic feet
cubic meters	1.308 cubic yards
liters	0.0013 cubic yards
square centimeters	0.155 square inches
square meters	10.764 square feet
square meters	1.196 square yards

U.S. and U.K. Units

Volume Conversions

Unit	Unit
U.S. gallons	0.833 U.K. gallons
U.S. gallons	3.785 liters
U.S. quarts	0.946 liters
U.S. ounces	0.029 liters
U.K. gallons	1.2 U.S. gallons
U.K. gallons	4.55 liters
U.K. quarts	1.136 liters
U.K. ounces	0.028 liters
liters	0.264 U.S. gallons
liters	0.22 U.S. quarts
liters	1.056 U.K. gallons
liters	0.879 U.K. quarts

Fahrenheit and Celsius

USDA Hardiness Zone Temperature Conversions

Planting Zone	Fahrenheit	Celsius
1	Below –50°	Below –46°
2	–50° to –40°	–46° to –40°
3	–40° to –30°	–40° to –35°
4	–30° to –20°	–35° to –29°
5	–20° to –10°	–29° to –23°
6	–10° to 0°	–23° to –18°
7	0° to 10°	–18° to –12°
8	10° to 20°	–12° to –7°
9	20° to 30°	–7° to –1°
10	30° to 40°	–1° to 4°
11	Above 40°	Above 4°

Temperature Conversions
(Degrees Fahrenheit—32) x 5/9 = Degrees Celsius
(Degrees Celsius x 1.80) + 32 = Degrees Fahrenheit

Resources

BOOKS

Ballou, Burt, ed. *AKCA Guide to Koi Health*. Associated Koi Clubs of America, 2000.

Barber, Terry Anne. *Waterfalls and Fountains for Your Garden Pond*. TFH, 2003.

Blasiola, George C. *Koi: A Complete Pet Owner's Manual*. Barron's, 2005.

Bridgewater, Alan, and Gill Bridgewater. *Outdoor Stonework: 16 Easy-to-Build Projects for Your Yard and Garden*. Storey Communications, 2001.

Cole, Peter. *The Art of Koi Keeping: A Complete Guide*. Blandford, 1995.

Davitt, Keith. *Water Features for Small Gardens: From Concept to Construction*. Timber Press, 2003.

Dimmock, Charlie. *Ground Force Water Garden Workbook*. BBC Worldwide, 1999.

Fisher, Kathleen, *Complete Guide to Water Gardens*. Creative Homeowner, 2000.

Hickling, Steve, ed. *Koi: Living Jewels of the Orient*. Barron's, 2002.

Hirst, Bryan. *Building Garden Ponds: 10 Step-by-Step Projects*. Voyageur Press, 2004.

Muha, Laura, and Jeffery Kurtz. *Landscaping Your Garden Pond*. TFH, 2004.

Nash, Helen. *Country Living Gardener's Water Features for Every Garden*. Hearst Books, 2000.

Rees, Yvonne, and Peter May. *The Water Garden Design Book*. Barron's, 2001.

Robinson, Peter. *The American Horticultural Society Complete Guide to Water Gardening*. DK Publishing, 1997.

Scott, Brian M. *The Super Simple Guide to Koi*. TFH, 2003.

Slocum, Perry D. *Waterlilies and Lotuses: Species, Cultivars, and New Hybrids*. Timber Press, 2005.

Speichert, Greg, and Sue Speichert. *Encyclopedia of Water Garden Plants*. Timber Press, 2004.

Swindells, Phillip. *The Master Book of the Water Garden*. Bullfinch, 2002.

Thomas, Charles M., and Richard M. Koogle. *Ortho's All About Building Waterfalls, Pools, and Streams*. Meredith Books, 2002.

MAGAZINES

Koi World, published by BowTie Inc.

Koi USA, published by Associated Koi Clubs of America.

Ponds Magazine, published by BowTie Inc.

Ponds USA, published by BowTie Inc.

Water Gardening Magazine, published by
The Water Gardener's Inc.

WEB SITES

Organizations

Associated Koi Clubs of America
http://www.akca.org

International Waterlily and Water Gardening Society
http://www.iwgs.org

North American Water Garden Society
http://www.nawgs.com

Victoria Adventure
http://www.victoria-adventure.org
Database of information on the cultivation of Victoria, waterlilies, and lotuses.

Professional Landscape Contractor Organizations

Professional Landcare Network (PLANET)
http://www.landcarenetwork.org
Formerly, the Associated Landscape Contractors of America, it includes a listing of certified U.S. landscape contractors.

Professional Pond Builder Organizations

Certified Aquascape Contractors
http://www.certifiedaquascapecontractor.com

International Professional Pond Contractor's Association
http://www.ippca.com

National Association of Pond Professionals
http://www.nationalpondpro.com

Glossary

Acidity A measure of the ability of a water to accept bases without the pH increasing.

Aerobic Processes that require oxygen (for example, the nitrogen cycle).

Algae Green- to brown-colored, chlorophyll-containing organism.

Alkalinity The ability of a solution to resist acidification.

Ammonia The major excretory product of fish and many other aquatic organisms.

Anaerobic Processes functioning in the absence of oxygen (for example, decay).

Biological filtration Using colonies of aerobic (nitrifying) bacteria to break down nitrogenous wastes (*see* nitrogen cycle).

Buffer A substance that can neutralize a base or an acid so that the original pH of the liquid is maintained or changes much more slowly than if the buffer were not present.

Carbon dioxide (CO_2) A colorless, odorless gas that readily dissolves in water to form carbonic acid. Carbon dioxide is respired by animals and absorbed by plants during photosynthesis.

Chemical filtration Nonbiological removal (for example, using activated carbon) of potentially harmful chemicals.

Chloramine A chemical like chlorine used to disinfect city water. More toxic to fish and more difficult to neutralize than chlorine.

Chlorine A chemical used to disinfect city or tap water. It is toxic to fish.

Compound A substance combining two or more elements from the periodic table of elements.

Dechlorinator A substance that is added to city water to get rid of chlorine and chloramines.

Detritus Dead organic material (either plant, animal, or bacterial) that can be degraded or mineralized by bacterial processes.

Gravity-fed multichamber filter A system wherein a pump carries water from the pond through a series of chambers with progressively denser filtration media, allowing waste to settle to the bottom of each chamber, before it returns, via gravity, to the pond.

Gravity-return filter A system that uses an in-pond pump to push the water to the filter chamber situated above the water surface and returns water to the pond under gravity.

Ground fault circuit interrupter (GFCI) An electrical safety device that shuts off the electrical current when it senses an electrical short.

Hardness A measure of the amount of calcium and magnesium in water expressed as calcium carbonate.

Head height A measurement of a pump's ability to push water over the rise and run of a section of piping, while being slowed by the friction of water on the inside of the pipe.

Ion Atom with an electrical charge; for example, when dissolved in water, the salt sodium chloride (NaCl) breaks (dissociates) into sodium (Na^+ and chlorine (Cl^-) ions.

Ion exchange A type of filtration whereby one compound (or element) changes place with another on the surface of a medium. Common uses for ion exchange media include removing hardness from water and ammonia removal by exchange also with sodium.

Magnetic appliances Pond appliances using motors that operate by means of a fluctuating electrical field, inside a sealed chamber that is not exposed to water.

Mechanical filtration The trapping of particulate material from pond water by straining the water through a pad, sponge, or similar medium, and the subsequent removal and cleaning of that medium.

Nitrate (NO_3) A form of nitrogen that is the end product of nitrification, which is produced by *Nitrospira* spp.

Nitrifying bacteria Aerobic bacteria that drive the nitrogen cycle.

Nitrite (NO_2) A form or nitrogen that is produced during nitrification and denitrification by bacteria.

Nitrogen cycle The breakdown of nitrogenous fish wastes into less toxic chemicals (ammonium to nitrite to nitrate) through the action of beneficial bacteria.

Ozonizers Mechanical units that release charged oxygen ions, which break down organic waste particles, into the water.

pH Measurement of the level of hydrogen ions in a solution on a scale from 1 (acidic) to 14 (alkaline or basic).

pH balance Refers to the process in which the oxidation processes in a pond (which tend to lower the pH) are counterbalanced by reduction processes (which tend to increase the pH), leading to an overall stable pH value.

Pressure filter A device that filters water under pressure through a series of progressive density media, thereby allowing clean water output at a distance from the device.

Pure water Water that contains nothing except a few dissociated hydrogen and hydroxyl ions. Pure water contains no salts, gases, bacteria, or other substances.

Skimmer Generally a box-shaped device positioned at the top edge of the pond and plumbed so that the pond surface water is drawn into the skimmer by a pump.

Symbiosis Partnership between two organisms; may take the form of mutualism, commensalism, or parasitism.

Ultraviolet (UV) clarifier A device that incorporates an ultraviolet lamp radiating light in the UV range that kills algae in the water column.

Undergravel filter A system of filtration that draws water through a layer of crushed stone or gravel, which houses beneficial bacteria that degrade toxins in the water.

Water quality A term that encompasses the entire range of physical, chemical, and biological environmental factors in a pond.

Index